T0137369

Project-Based Knowledge in Organizing Open Innovation

Sara Bonesso · Anna Comacchio
Claudio Pizzi
Editors

Project-Based Knowledge in Organizing Open Innovation

Editors
Sara Bonesso
Claudio Pizzi
Università Ca' Foscari Venezia
Venezia
Italy

Anna Comacchio
Venezia
Italy

ISBN 978-1-4471-7058-7 ISBN 978-1-4471-6509-5 (eBook)
DOI 10.1007/978-1-4471-6509-5
Springer London Heidelberg New York Dordrecht

Softcover reprint of the hardcover 1st edition 2014

Printed on acid-free paper

Springer is part of Springer Science+Business Media (www.springer.com)

Foreword

The discourse about projects, and project-based organizing, has gone a long way since its beginnings in the project management literature. It has generated vast attention and attracted considerable research efforts. Originally, project management developed out of the need to systematize activities undertaken to achieve specific goals, within a limited time period. The professionalization of project management activities, from the 1950s onwards, led to the development of a number of tools and techniques to plan, design, monitor, and implement tasks in such a way as to optimize along the axis of the dreaded "iron triangle" of costs, time, and quality. Most of the work done along these lines emphasized the uniqueness of project work, its distinctiveness from mass production, the difficulties of comparing contents and tasks across projects. Certainly, this discussion contributed greatly to the rise of the project manager as a new professional figure, with great status and legitimacy.

Over time though, scholars and practitioners have moved away from the analysis of specific projects. For example, Gann and Salter (2000) discussed the problem of exploring how project-level processes and objectives are related to business-level projects and objectives. Their departure from the traditional discussion of project management was grounded in the analysis of innovative projects in a variety of industries related to the so-called Complex Products and Systems area (e.g. complex civil engineering structures). In these industries, few projects are implemented, but each has a major and direct impact on a firm's performance. Hence, the need to understand how project-level decisions affects broader definitions of performance. In relation to this, a discussion has emerged about how firms which operate in a project-based environment can actually learn, over and above what is learnt within a project team. How do they transfer knowledge across projects? How do they build on lessons learnt without reinventing the wheel every time? What is the role of individuals in this process? And what instead can be codified via digital or physical means?

The discussion about organizational learning, memory, and knowledge codification merged with the literature on project-based activities. For example, Prencipe and Tell (2001) looked at different learning strategies deployed by firms in various industries; Cacciatori (2008) analyzed the role of knowledge codification in the form of objects and artifacts; Criscuolo et al. (2007) discussed the role of expert yellow pages' to codify and make visible the skills experts acquire while working on different projects.

Most of these papers build upon evidence collected in sectors which traditionally are organized by projects, like the construction industry or, more recently, the movie industry (see also Cattani et al. (2011) for a comprehensive review of issues and empirics). However, project-based organizing is also common in sectors that have normally been analyzed looking at the firm- or team-level. In fact, a remarkable gap in the vast literature on project-based organizing is the link with the wealth of research originating from the New Product Development (NPD) area, historically one of the earliest fields to grasp the analytical importance and empirical relevance of the project as a unit of analysis.

This book's first contribution is closing this important loop toward NPD research. The chapters touch upon issues related to NPD in a variety of sectors, some of which are not normally studied through the lens of project-based organizing, e.g., machine-tools and pharmaceuticals. In so doing, this book already delivers a first important evidence.

Second, this book is not only about closing loops, it is also about opening up toward new avenues of research. Great emphasis is given to the specificities of innovation processes which are increasingly open and distributed in nature, and the project-based nature of the environments in which they happen. How can a focus on projects advance the discussion in this direction? This book engages the audience by identifying a series of interconnected questions that, taken together, provide an interesting, comprehensive, and progressive view of how organizations in a variety of industries can strategize by leveraging the peculiarities of project-based organizing, rather than organizing in order to overcome its limitations. In a way, the chapters collected in this book turn the story upside down. It is not that organizations persist despite the intermittent nature of their projects, which present them with challenges that need to be overcome through knowledge management, memory tools, yellow pages, etc.; rather, it is the opposite: projects become the tools through which organizations adapt, adjust and respond to changing environmental circumstances.

How can projects be leveraged as strategic tools to engage external and internal stakeholders? How can organizations become selectively open and closed? How can they optimize internal and external sourcing strategies? How can they reconcile the traditional tradeoff between exploitation and exploration? How do project-based activities interact with firm-level strategies, such as R&D alliances? This is but a sample of the tremendously important questions raised in this book, which will soon become a must read for the scholarly community interested in project-based activities.

Zurich, October 2013 Stefano Brusoni

References

Cacciatori, E., (2008). Memory objects in project environments: storing, retrieving and adapting learning in project-based firms. *Research Policy*, 37(9), 1591–1601.

Cattani, G., Ferriani, S., Frederiksen, L. and Täube F., (2011), Project-based organizing and strategic management: a long-term research agenda on temporary organizational forms. In Cattani, G., Ferriani, S., Frederiksen, L., Täube, F., (ed.) Project-Based Organizing and Strategic Management (Advances in Strategic Management, Volume 28), Howard House, Wagon Lake, UK: Emerald Group Publishing.

Criscuolo, P., Salter, A., Sheehan, T., (2007). Making knowledge visible: Using expert yellow pages to map capabilities in professional services firms. *Research Policy*, 36(10), 1603–1619.

Gann, D. M., Salter, A. J., (2000). Innovation in project-based, service-enhanced firms: the construction of complex products and systems. *Research Policy*, 29(78), 955–972.

Prencipe, A., Tell, F., (2001). Inter-project learning: Processes and outcomes of knowledge codification in project-based firms. *Research Policy*, 30(9), 1373–1394.

Acknowledgments

We would like to take this opportunity to thank a number of people who have contributed to this book. We are grateful to our colleagues of the Department of Management and the Department of Economics of Ca' Foscari University of Venezia for the support and insightful comments provided. More specifically, we want to mention the people who generously gave their time and shared their experience with us during the different stages of our writing process.

We would like to say a special thank you to our colleague Maria Rees for her technical advice and guidance that allowed us to further improve the editing of this book.

We also wish to thank Stefano Brusoni for his insightful Foreword to this book.

Contents

Contributors

Markus Becker Strategic Organization Design Unit, University of Southern Denmark, Odense M, Denmark

Sara Bonesso Department of Management, Ca' Foscari University of Venezia, Venezia, Italy

Anna Comacchio Department of Management, Ca' Foscari University of Venezia, Venezia, Italy

Luisa Errichiello Institute for Service Industry Research (IRAT), Italian National Research Council (CNR), Napoli, Italy

Claudio Pizzi Department of Economics, Ca' Foscari University of Venezia, Venezia, Italy

Giulia Trombini Department of Management, Ca' Foscari University of Venezia, Venezia, Italy

Francesco Zirpoli Department of Management, Ca' Foscari University of Venezia, Venezia, Italy

Acronyms

B2B	Business-to-Business
CAM	Computer Aided Manufacturing
CoPS	Complex Product Systems
IPC	International Patent Classification
IT	Information Technology
M&A	Mergers and Acquisitions
NBER	National Bureau of Economic Research
NPD	New Product Development
ODM	Original Design Manufacturer
OEMs	Original Equipment Manufacturers
OLS	Ordinary Least Squares
R&D	Research and Development
SIC	Standard Industrial Classification
UCIMU	Association of Italian manufacturers of machine tools, robots, automation and ancillary products
VAR	Value-Added Reseller
VIF	Variance Inflation Factor

Chapter 1
Leveraging on Projects to Strategically Organize Open Innovation

Sara Bonesso and Anna Comacchio

Abstract This chapter proposes a project-based view of open innovation in contexts where knowledge is dispersed and the locus of innovation does not reside inside a single organization. After a review of studies on the organization of innovative labor and the role of new product development projects, the chapter discusses how innovative projects could be conceived as a strategic site where the organization of the external network of knowledge sources takes place. It analyzes the main factors explaining firms' propensity to design open innovation project-by-project. Adopting a contingency approach, it reviews recent research on the project features affecting inbound choices in innovative organizations and on the relation between project choices and the knowledge base of the firm. Finally, the chapter provides an overview of the chapters of the book and how they address some key theoretical and empirical open issues.

1.1 Studying the Organization of Innovative Labor: The Locus of Innovation and the Role of Innovative Projects

Innovation is a complex endeavor at the core of a firm's competitive advantage, and how to organize it is a central issue for managers and a key topic in studies of organization, strategy, and innovation management. In several sectors where knowledge expertise is dispersed, the locus of innovation does not reside inside a single organization but can be found in a network of specialized external knowledge sources (Powell et al. 1996). Thus, the endeavor of targeting a new market with a novel

S. Bonesso (✉) · A. Comacchio
Department of Management, Ca' Foscari University of Venezia,
San Giobbe Cannaregio, 873, 30121 Venezia, Italy
e-mail: bonesso@unive.it

A. Comacchio
e-mail: comacchio@unive.it

S. Bonesso et al. (eds.), *Project-Based Knowledge in Organizing Open Innovation*,
DOI: 10.1007/978-1-4471-6509-5_1, © Springer-Verlag London 2014

product or service idea implies organizational choices on how to divide innovative labor among partners and which pathways to and from the external market of ideas are to be strategically deployed.

While there is a consensus on the fact that innovation strategies are implemented through multiple projects (Gemünden et al. 2013; Wheelwright and Clark 1995), less attention has been devoted to strategic organizing through projects and to what role a set of projects plays when a firm has to choose how to organize innovative tasks across organizational boundaries.

Traditionally, scholars, considering an internal locus of innovation, analyzed how "different kinds of innovation require different kinds of organizational environments and different managerial skills" (Abernathy and Clark 1985). In this stream of research, innovation projects played a role in understanding the organization of innovation. The new product development (NPD) projects were studied as microcosmos of the whole organization (Clark 1995) and a core organizational unit through which innovation is organized within innovative firms in the automotive, aerospace, and internet-software industries (Brusoni et al. 1998, 2001; Clark 1995; MacCormack et al. 2001). Research on product development projects investigated the internal organizational solutions for innovative labor division and task allocation, which reduce time to market, increase creativity, and foster the core firm's innovation capabilities (Wheelwright and Clark 1995). For instance, role differentiation within a project team has been studied as a mechanism to balance the external search for knowledge with efficient project management (Chakrabarti and Hauschildt 1989), and concurrent engineering was identified as a way to facilitate inter-unit task allocation and coordination by anticipating problems with manufacturing during the early stages of an NPD project (Jurgens 1999).

The debate on the internal organization design of innovative activities has been enriched by increasing attention to task decomposition and the efficient design of firm boundaries. In the last few decades, there has been considerable growth in collaborative partnerships and increase in their variety, with a more recent substitution of long-standing partnerships with more flexible and blended ones (Grant and Baden-Fuller 2004; Hagedoorn et al. 2008; Roijakkers and Hagedoorn 2006; Trombini and Comacchio 2012).

The innovation management literature shifted the attention from intra to inter-organization labor division (Arora and Gambardella 1994; Arora et al. 1997; Powell et al. 2004), and showed that inputs to innovation could be organized across firms' boundaries, opening up product development to a network of partners through task decomposition and modularity (Langlois and Robertson 1992). This allows for the exploitation of different loci of innovation, with the consequent co-specialization of organizations along the innovation process, through new forms of distributed innovation. This is the case of the pharmaceutical sector, where small biotech firms and big pharma companies collaborate with different roles in the stages of a new compound project.

The analysis of the efficiency drivers of the division of innovative labor was deepened by studies adopting the transaction cost perspective. This stream of research mainly considers a firm as a whole and analyzes boundary positioning. The technology sourcing agreement design (from licensing to equity partnerships) a firm adopts is aimed at easing knowledge transfer and mitigating partners' opportunistic behaviors (Colombo 2003; Oxley 1997).

The knowledge-based view turns attention to processes of knowledge accumulation, transfer, and combination to explain the internal and external division of innovative labor (Grandori and Kogut 2002; Kogut and Zander 1992). For instance, it analyzes how to locate internal R&D activities between corporate and divisional units (Tidd et al. 2005), or how to geographically decentralize R&D in global companies (Zedtwitz and Gassmann 2002) to build a responsive organization. This stream of research also points to the formation of external agreements, looking at antecedents and benefits in accessing knowledge sources that are complementary to a firm's knowledge base (Grant and Baden-Fuller 2004). Notwithstanding the concern for knowledge combination processes, the role of innovation projects and of the project level of analysis is underestimated in this literature. This stream of research focuses instead on the firm level of analysis, in order to test the core research question of whether and how a firm's and its partners' knowledge base features such as breadth and depth, cognitive distance (Nooteboom et al. 2007), and absorptive capacity (Cohen and Levinthal 1990) are predictors of organizational boundary choices.

In this debate, a different perspective was taken by studies on complex product systems and system integrators (Brusoni et al. 1998; Prencipe 2003). The more fine-grained lens of the project-based organizations was adopted to investigate how innovative labor is divided among firms in sectors such as the aerospace or automotive industries to exploit assets and economies of scales of component producers, on the one hand, and proximity to customers' needs and control on architectural knowledge of a "focal" firm, on the other. These studies puts forward research on labor division for innovation, pointing to the importance of the knowledge base of the firm, and the relationship between specialization in knowledge production and the boundaries of the firm (Brusoni et al. 2001).

A renewed attention to projects has recently been put forward by the open innovation literature (West et al. 2014). It aims at investigating how firms increase the range of knowledge recombination opportunities by tapping a large system of distributed and differentiated partners, including suppliers, customers (Von Hippel 2005), research institutions, and intermediaries (Howells 2006). Contrasting the closed approach with a distributed one, the open innovation model (Chesbrough 2003, 2006) studies how a firm, by opening up organizational boundaries, could increase value added. Empirical evidence shows the diffusion of the open innovation principles among companies in different sectors, from high-tech to more mature ones (Chiaroni et al. 2011; Gassmann et al. 2010) of different sizes, from large to small-medium companies (Vrande et al. 2009) and the implications of this model on a firm's performance (Rass et al. 2013).

In this context, projects are studied from a process perspective, by a stage-gate approach, suggesting that "in an open innovation process, projects can be launched from internal or external sources and new technology can enter at various stages" (Elmquist et al. 2009, p. 327). Recent research shows that opportunities of opening up the boundaries with outbound and inbound knowledge flows are different at distinct stages of a project development process (Grönlund et al. 2010).

Besides the contribution on the benefits and the risks of importing and exporting know-how at different project stages, open innovation research has provided little empirical evidence on "How to design open innovation" (Christiansen et al. 2013; Huizing 2011). To overcome this gap, a contingency perspective of open innovation has recently emerged, exploring questions related to factors that affect the conditions under which open innovation is effective, and suggesting more specifically that the debate should face three open issues: (1) the relationship between type of "openness" and type of "innovation", (2) when open innovation is timely in respect to the product life cycle, and (3) the degree of openness a firm should pursue (Elmquist et al. 2009).

In summary, the book aims to raise the question about the relevance of projects and the project level of analysis in studying the organization of innovative activities in contexts where innovation is open and distributed. In this section, we highlighted the relationship between projects and the organization design of in-house innovation activities. However, as Gulati et al. recently maintained, "an emphasis on intra-firm design may be out of date or, at the very least, incomplete" (Gulati et al. 2012, p. 572). Notwithstanding the growing body of research on the division of innovative labor between firms, both transaction-cost and knowledge-based perspectives confined their main focus on a firm's level of analysis. Scholars of both perspectives were less concerned about exploring at micro level when and how firms might "expand their boundaries", while controlling their core innovation capabilities (Gulati and Kletter 2005).

We propose that the micro-level analysis of NPD projects helps to tackle the previous literature gaps in the innovative labor division in temporary and permanent companies where innovation is primarily organized by projects, and the relationship between the degree of openness of a firm and its project-portfolio features is relevant (Lakhani and Tushman 2012; Mortara and Minshall 2011).

The adoption of a micro-perspective on projects could fill this gap and provide a novel understanding of how permanent firms organize innovative labor across organizational boundaries.

1.2 Project-Based Open Innovation: Innovative Projects as a Strategic Form of Organizing

The question of how a set of discrete and diverse NPD projects is related to a company's open innovation solutions challenges the idea of a firm's single strategy. It suggests that for some firms it is more appropriate to take a project-based view of open innovation.

According to the discussion in the previous section, the conceptual and empirical gaps can be addressed along two main dimensions: *time and temporality* in open innovation solutions and *degree of openness* of a organization, namely the scope of multiple boundary solutions a firm can simultaneously implement project-by-project. We suggest that these two dimensions are central not only in the recent debate on open innovation, but also in two different but complementary streams of research: project portfolio management and studies on project-based or temporary organizations. Bridging these different studies would increase our understanding of how firms might open their boundaries on a project basis.

Temporality is central to the recent debate on project-based organizations (Bakker 2010), which analyzes the temporary organizations, defining them as "a set of organizational actors working together on a complex task over a limited period of time" (Bakker 2010, p. 468). Indeed, project-based organizations are spreading in different sectors and across different types of firms (Bosch-Sijtsema and Postma 2009; Hobday 2000). Due to their temporary nature, projects are a key organizational mechanism for pursuing innovation even in permanent organizations. Projects are "transitory organizational configurations" and are flexible tools that allow a firm to "modify direction speedily as knowledge and markets evolve" (Cattani et al. 2011, p. xix). Thus, we suggest that firms, operating in volatile markets or in sectors characterized by rapid technological changes, might choose a project-based open innovation approach to benefit from the inherent flexibility of project organizing. Temporality, however, has drawbacks concerning knowledge transfer and accumulation. Firms that organize innovation within and across boundaries through projects face the problem of how to transfer the learning-by-doing occurring at a project level to other projects as well as at the organizational level of analysis (Cattani et al. 2011). Thus, the short-term plasticity of a project portfolio helps to face external technological and competitive threats and opportunities. However, it has to be balanced by the focus on long-term knowledge accumulation at the firm level (Brusoni et al. 2001; Nooteboom et al. 2007).

As far as *degree of openness* is concerned, the literature on innovation and project management shows that a firm's innovation strategy is constituted by a portfolio composed of projects with different degrees of innovativeness (platform and derivative projects), at different stages of development (Battistella and Nonino 2012; Gassmann et al. 2010; Grönlund et al. 2010). The different degree of innovativeness of a project impacts on the boundary choices of a firm in terms of task decomposition and range of external sources (Cattani et al. 2011; Gulati et al. 2012).

A few studies have investigated the antecedents of a simultaneous set of boundary options in a firm. Addressing this literature gap, scholars on open innovation have recently devoted their attention to how organizations, adopting an open innovation approach, deal with a wide range of decisions regarding external parties according to different project features (Bahemia and Squire 2010). Consequently, the scope of boundary solutions varies not only across firms, as argued in the open innovation literature, but within a single organization, and research should investigate more in-depth the extent to which each single project requires external knowledge sources.

In summary, the debate on project portfolio management for innovation as well as the more recent debate on project-based organizations have contributed to raise the attention on the centrality of projects in our understanding of how firms organize innovation. However, less research has been devoted to how firms, managing project-by-project, build an open innovation strategy. We propose that a project-based view on open innovation bridges separate but complementary themes of research, opening new avenues of research into the role of multiple projects in designing multiple cooperative relationships. We suggest that the drivers of the firm's project-based open innovation approach could be analyzed along two main dimensions: the benefits and disadvantages of flexibly exploiting the temporality of projects and the degree of openness of a firm's project porfolio.

1.3 The Project Level of Analysis in Inbound Innovation Processes

As discussed in the previous sections, the project is the organizational building block in temporary firms where it is the primary unit for production, organization, and innovation (Cattani et al. 2011), and it is increasingly a key organizational solution through which permanent organizations flexibly develop innovation.

Our argument is that projects are powerful exploratory tools for both the knowledge content and knowledge sources landscape. As maintained by Lenfle (2008, p. 471) "the result of the project is then no longer simply a product," but the opportunity of learning new knowledge that can significantly foster the capacity to be innovative (Söderlund et al. 2008). First, the problem-solving process, involved by an NPD project, may boost a simultaneous search for innovative knowledge solutions and external sources (Rothaermel and Deeds 2004). A firm facing a new product development problem grounds on project requirements the search for the external partner that offer the best matching solution. The in-depth knowledge of the project helps to enlarge the search scope beyond the knowledge related to the product itself, including the partner's complementary capabilities for the product development process. This is particularly true for non-incremental NPD projects that do not merely reuse extant knowledge. A recent study on a set of breakthrough NPD projects showed that the process, undertaken at project level, of abstracting the problem by in-depth technical and contextual analysis helps to search outside through analogical thinking and recognize highly novel solutions. This type of search increases the chance of finding non-obvious solutions in distant sectors, raising consequently the probability to rely on novel partners (Gassmann and Zeschky 2008). Non-incremental NPD projects prompt the search for new knowledge sources outside the organizational boundaries (Rothaermel and Deeds 2004). Thus, while some firms rely on a uniform pattern of technology sourcing across different NPD projects, others may choose an inbound configuration on a project-by-project basis.

Second, firms, to actively search and timely exploit new avenues of research or technological opportunities, are spurred to choose partners project-by-project, widening the scope of sourcing choices over time. The literature implicitly assumes that once a firm has adopted a sourcing strategy, it is locked in the set of relationships it has built in the past. By contrast, research in the last decade shows that the need for strategic flexibility has prompted firms to move from long-term decisions to a more short-term and contingent approach to joint knowledge development (Dittrich and Duysters 2007; Hagedoorn 2002; Knudsen and Mortensen 2011). Therefore, the need to flexibly and quickly redesign the external network, according to fast-changing technology and competition, pushes firms to leverage on the flexibility of projects and consider a project-by-project open innovation approach over time (Bosch-Sijtsema and Postma 2009; Hobday 2000; Lenfle 2008).

Third, as discussed in the previous section, the diffusion of open innovation among firms could be described as a continuum between firms with a higher degree of openness and firms adopting a closed R&D model, and we suggest that this continuum can also be applied to the different projects of a single firm. A company achieving divergent innovative aims by a diversified NPD project portfolio might coherently design the external collaborations of each project on a continuum from a high to a low degree of engagement. This means that firms might simultaneously adopt multiple boundary options on a project basis. By choosing strategically to integrate different degrees of openness and innovativeness at project portfolio level (Knudsen and Mortensen 2011), a firm can exploit a wider range of benefits and mitigate the diminishing returns from open innovation (Laursen and Salter 2006; Leiponen and Helfat 2010).

In summary, in this paragraph we argued that there are three main reasons for a firm to choose a project-by-project division of innovative labor and task allocation among a network of external partners and to couple knowledge and partner search: (1) knowledge to be developed by an NPD project is the relevant input for a joint search for novel and distant knowledge and for adeguate partners; (2) a project-by-project division of innovative labor makes it possible to flexibly adapt the network of partners to new technological and competitive opportunities over time; (3) a project-by-project approach makes it possible to simultaneously exploit the benefits of different degrees of openness while mitigating related risks and costs.

Accordingly, we suggest that the adoption of the project level of analysis to study inbound sourcing enhances our understanding of a company innovation and organization strategy as informed by a project-portfolio oversight, deepening our investigation into the organization of innovation as a set of simultaneous diverse boundary options (Bahemia and Squire 2010; Knudsen and Mortensen 2011).

Coherently with these arguments, in the following section we analyze the project features that could affect a boundary option.

1.4 Project Features as Contingent Factors in Inbound Open Innovation

In the previous section, we discussed the advantages that the project level may yield to the analysis of how firms organize inbound open innovation. Such a fine-grained level of analysis is salient especially when the characteristics of each NPD project vary substantially within the same firm, demanding different degrees of openness.

It is only recently that scholars have criticized the research that considers open innovation exclusively as a firm-level phenomenon and have claimed that "the decision for managers is not solely a binary choice of whether to adopt open innovation at a firm level" (Bahemia and Squire 2010, p. 604).

Adopting a micro approach makes it possible to theorize explanatory antecedents that are located at the same level of analysis of the phenomenon to be explained. Nevertheless, the majority of the studies collects data about antecedents of openness at the firm level or consider only the most important NPD project carried out by the company (Huang et al. 2009; Knudsen and Mortensen 2011; Tranekjer and Søndergaard 2013), rather than focusing on each single project's features that may lead firms to define different degrees of openness of their innovation processes. Indeed, in the absence of project-level data, studies have considered some of the company's attributes as determinants, for instance the volatility and ambiguity of a company's environment (Carson et al. 2006), R&D intensity (Mol 2005), breadth of a firm's knowledge stock (Zhang et al. 2007), breadth and depth of technological capabilities (Prencipe 2000), and absorptive capacity (Tsai 2009). These studies examine the endowment of a company as explanation of inbound decisions, considering for instance how past accumulation of R&D knowledge can be a premise that allows a firm to recognize and absorb the knowledge of external partners (Cohen and Levinthal 1990; Nooteboom et al. 2007).

The project management literature points to projects as a knowledge management tool since, within the firm, they provide a space for learning and knowledge creation (Nilsen 2013). However, these contributions, focusing primarily on the internal learning processes, neglect the role of the knowledge features of NPD projects in explaining why and how firms decide to span their boundaries in search of new ideas and partners. On the other hand, the relevance of contingencies in influencing project-related decisions has been highlighted by a recent bibliometric study on the use of contingency theory in project management research (Hanisch and Wald 2012). The findings point to relevant factors such as project types (construction projects, IT development projects, manufacturing projects, and R&D projects) and social factors (culture, leadership, relationships, and teams), but the role of knowledge-based contingent factors, like the degree of novelty or basicness of the projects, is neglected. This gap in the project management literature can be addressed by bridging this stream of research with the open innovation studies, that have only very recently started to devote attention to the project level of analysis (West et al. 2014). A summary of the few recent articles that, adopting a project-level contingency approach, add to this line of research is provided in Table 1.1.

For each study, the knowledge attributes are defined at project level along with their impact on the degree of openness. Project novelty emerges as the most frequent knowledge dimension investigated in these studies, followed by the degree of tacitness and basicness of the new developed knowledge as well as breadth and strategic importance. The table shows the impact that different degrees of these attributes have on the inbound decisions, such as internal-versus-external sourcing, mode of

Table 1.1 Project knowledge dimensions and open innovation

Reference	Sample characteristics	Project's knowledge features	Definition	Impact on inbound open innovation
		Basicness	The extent to which a project refers to fundamental research and is aimed at developing new knowledge as opposed to exploiting previously held knowledge	The more basic a project, the more the firm relies on universities and other firms through cooperative arrangements
		Novelty	Extent to which a project's knowledge is novel relative to the firm's existing knowledge base	The more novel a project: • The more the firm relies on universities through contracting • The less the firm relies on other firms through cooperative agreements
Cassiman et al. (2010)	52 projects carried out by a multinational company in the semiconductor industry			
		Strategic importance	The extent to which the firm will rely on the knowledge involved in the project to build its competitive advantage	The more a project leads to strategic results: • The more the firm relies on universities through contracting • The less the firm relies on other firms through cooperative agreements
		Codifiability	The degree to which knowledge can be codified as opposed to being tacit	The higher the codifiability of the project, the more the firm relies on universities and other firms through cooperation agreements and contracting

(continued)

Table 1.1 (continued)

Reference	Sample characteristics	Project's knowledge features	Definition	Impact on inbound open innovation
Bonesso et al. (2011)	60 NPD projects carried out by seven companies operating in the machine tools industry	Knowledge novelty	Degree to which a project provides superior product functionalities for customers	The higher the knowledge novelty in the NPD project, the more likely it is that the firm will rely on external sourcing
		Knowledge breadth	Degree of diversity or heterogeneity among technological domains a project draws on to solve different primary problems	The higher the knowledge novelty in the NPD project, the more likely it is that the firm will rely on external sourcing
Hsieh and Tidd (2012)	4 NPD projects, case study in firms operating in the convenience chain stores	Project novelty	Newness of components or functions	The more novel development projects demand higher levels of interaction with existing suppliers to communicate. The more novel development projects are associated with interaction with additional suppliers and new partners to expand the scope and function of new services
Salge et al. (2013)	62 NPD projects initiated within public health-care organizations of the English National Health Service	Project type (exploitative NPD projects and explorative NPD projects)	Exploitative projects improve an already existing product, service, or process Explorative projects create an entirely novel product, service, or process	The positive effect of search openness on new product creativity and new product success are stronger for explorative NPD projects.

governance (cooperation agreements vs contracting), and type of external sources (universities and research labs vs. supply chain partners).

The new insights provided by the contributions summarized in Table 1.1, the adoption of a project-level contingency approach may be fruitful both to open innovation research and to strategic project management literature in order to address still open issues. First, the impact of project knowledge features may be investigated considering the complex set of inbound choices (internal versus external sources, organizational forms, partner selection). Second, the effect of the project knowledge attributes on boundaries decisions may be contingent to the different stages of the NPD process. Third, research should devote attention to the role played by the industry in moderating the relationship between project features and boundary choices. Finally, as discussed above, the extant studies adopt the firm level of analysis or the project one, rather than carrying out a multi-level research on the relationship between project knowledge features and the knowledge base of the firm and how this relationship affects inbound choices.

1.5 Reconciling Project-Knowledge Dimensions and the Firm's Knowledge Base

The knowledge-based approach has recently explained the benefits of an integration among multiple levels of analysis, in order to better understand the processes and the mechanisms that can facilitate the transfer of knowledge across levels (individual, collective and organizational) within a firm (Foss et al. 2010). However, this internal knowledge-sharing focus has overshadowed the relevant issue of how external knowledge can be acquired and transferred across multiple levels. The previous section points to the need to identify the relevant project-knowledge features that may have an impact on inbound open innovation decisions. According to the recent call to study phenomena from a multi-level perspective, the investigation of how firms strategically organize open innovation may benefit from the integration of the project-level contingency approach and the firm-knowledge-based research.

The concept of a firm's knowledge stock or accumulation can be conceived both in terms of technological and organizational capabilities (Reich et al. 2012). As far as the firm's technological knowledge dimensions are concerned, the previous section explained how prior research has mainly considered features such as the breadth and depth of a firm's knowledge stock in explaining the propensity of a firm to rely on external sources. Concerning the organizational capabilities that may influence the open innovation decision-making process, a central role is played by the level of a firm's absorptive capacity (Cohen and Levinthal 1990). The literature has defined absorptive capacity as a precondition of inbound open innovation (Chiaroni et al. 2010), and research has recently devoted attention to how firms may build organizational practices that allow them to identify new and highly valuable product opportunities, assimilate internally the acquired external knowledge, and utilize it for innovation purposes (Lewin et al. 2011). Moreover, studies address the issue of how project teams may develop a knowledge integration capability in order to convert single members' knowledge resources into a collective one. Research shows that firm peformance is related to the adoption of key mechanisms that support the knowledge transfer across levels, such as transactive memory systems, information pooling, and functional diversity (Gardner et al. 2012).

Despite the insights provided by the literature that addresses the issue of organizing open innovation at the firm or at project level, research has remained silent on the possible integration between the two levels, and specifically two open issues remain unanswered. The first concerns how the two levels of analysis interact in determining inbound open innovation choices, and more specifically how the technological knowledge base of the firm influences or moderates the relationships between the project-knowledge features and the open innovation choices. The second issue is related to the role of the mechanisms through which firms can reconcile the acquisition of external knowledge at the project level with knowledge accumulation at the corporate one. As highlighted by a recent study, much of the knowledge generated in NPD projects is tacit, therefore it is difficult to express and is dependent on the interactions within project teams (Goffin and Koners 2011). This is especially the case of complex product development, where product architectures are integral and comprise a large number of components and subsystems with many technical interdependencies between them (Brusoni et al. 2001; Prencipe 2003; Takeishi and Fujimoto 2003). The multi-technology and multi-component nature of these products requires firms to rely on external partners that provide specialized knowledge which is often "tacit and can be transferred to the NPD project team through time-consuming learning-by-doing in close cooperation with the supplier" (Salge et al. 2013, p. 662). Therefore, the strategic organization of open and distributed innovation processes in the case of complex product development implies two main decisions. On the one hand, firms have to define to what extent they allocate innovation tasks to external specialized partners. On the other hand, they have to decide how to integrate and coordinate at firm level the external knowledge acquired by each single NPD project.

1.6 Conclusions

The book aims to extend the current theoretical debate on open innovation by focusing on the project level of analysis.

Chapter 2 "Exploring the knowledge space through project-based sourcing" adopts a project-level contingency approach for explaining inbound sourcing choices. It defines New Product Development (NPD) project as a strategic means used by innovative firms to explore the knowledge space for high-value solutions, and to search for external sources. The authors suggest that the knowledge space explored by an NPD project is grounded on the main elements of an industrial innovation system and is characterized by two key dimensions, namely knowledge novelty, the knowledge space of the performance features of a product that meet new customer needs, and knowledge breadth, the knowledge space of technological domains to draw on for solving product-related problems. The research is carried out on a sample of NPD projects developed by a group of leading firms operating in the machine tools industry. The chapter investigates, in companies that define sourcing on a project-by-project basis, the impact of the two project-knowledge features on the propensity to rely on external sources, to choose R&D development agreements as the governance form to involve partners, and to search cognitive distant partners instead of similar ones. Proposing a project-based approach to strategically organize inbound sourcing, the chapter provides evidence on the concept of a company sourcing strategy as a portfolio of decisions across projects.

Chapter 3 "A project-based perspective on complex product development" contributes to the debate concerning the organization of new product development in the case of complex products and specifically addresses the issue of reconciling the firm and the project levels of analysis. The authors focus on the challenges of leveraging on external distributed knowledge, pointing to the specific problems brought by the crucial role of "learning-by-doing" in complex product innovation processes. The authors argue that NPD performance is generated at the project level and it is here that the system integrator must be able to mobilize its integration competences. The tacitness of at least some knowledge, required to integrate systems, poses knowledge transfer problems among projects and between a project and the firm knowledge base. The chapter suggests that future studies should look at the powerful consequences of providing opportunities for learning-by-doing at the project level and argues that this learning process is crucial for the capability of integrating systems in projects.

Chapter 4 "Analysis of in-licensing decisions at a project and firm level: evidence from the biopharmaceutical industry" investigates how firms organize license-ties in order to reconcile resource access at a project level with knowledge accumulation dynamics at corporate level. The chapter analyzes what guides firms in choosing the governance modes through which to acquiring external innovations. Specifically, it focuses on the choice of combining or not the license with an R&D collaboration agreement. The author contends that the choice should be investigated considering two levels of analysis: the project level and the company level. The organization of license-ties is dependent on the features of the underlying licensed technology as well as on the structure of the firm's knowledge base. The chapter provides empirical evidence on how the two levels of analysis create contingencies with one another and determine how firms blend project-resource access with corporate knowledge accumulation dynamics.

Finally, Chap. 5 "Open innovation at firm and project level: future research agenda" discusses the contributions provided by the book. The main results and implications are compared and analyzed in light of the open issues raised in the present chapter. Converging results and key theoretical issues are presented and future lines of research are proposed. This final chapter provides a comparative analysis of the main open issues discussed in the previous chapters (Table 1.2).

Table 1.2 Links between chapters and open issues

Open issues	Chapter 2 Bonesso, Comacchio and Pizzi	Chapter 3 Becker, Errichiello and Zirpoli	Chapter 4 Trombini
Project versus firm level of analysis and methods	Project Quantitative study Cross-sectional survey 60 NPD projects	Project and firm Theoretical discussion	Project and firm Quantitative study Cross sectional dataset (1985–2004) 186 license projects
Knowledge type and features as antecedents of inbound choices	Technological knowledge: • Knowledge novelty • Knowledge breadth	Organizational knowledge • Knowledge on the integration of components and systems	Technological knowledge: • Distance of the in-licensed project from the firm's knowledge base • Depth of firm's knowledge base • Breadth of firm's knowledge base
Definition of inbound open innovation choices	• Internal-versus-external sourcing • R&D development agreements • Partner distance	Concurrent sourcing	License-project combined with an R&D collaboration and stand-alone licensing

First, concerning the level of analysis adopted in investigating the inbound choices, Chap. 2 points to the project-level contingency approach, whereas Chaps. 3 and 4 aim to reconcile both project and firm level of analysis.

Second, as regards the characteristics of the knowledge features as relevant antecedents of inbound choices, Chap. 2 focuses on knowledge novelty and breadth providing a fine-grained conceptualization and operationalization of these two dimensions at project level. Chapter 3 draws attention to knowledge about technical interdependencies existing among components and subsystems and about technologies involved in the component development. Chapter 4 considers the impact of the distance of the in-licensed project from the firm's knowledge base and includes as moderating variables the depth and breadth of the firm's knowledge base.

Third, as far as the inbound choices are concerned, Chap. 2 considers a set of three main decisions: internal-versus-external sourcing, the use of R&D development agreements as mode of governance, and the degree of cognitive distance with the partner. Chapter 3 analyzes the phenomenon of concurrent sourcing, and how firms simultaneously use make and buy options. Chapter 4 focuses on the mode of governance of inbound open innovation, and specifically on the decision about whether or not to combine license-project with an R&D collaboration.

Finally, the three chapters provide empirical and theoretical contributions in advancing the understanding on the research gaps.

References

Abernathy, W.J., Clark, K.B., (1985). Innovation: Mapping the winds of creative destruction. *Research Policy* 14(1): 3–22.

Arora, A., Gambardella, A., (1994). Evaluating Technological Information and Utilizing it. *Journal of Economic Behavior and Organization* 24(1): 91–114.

Arora, A., Gambardella, A., Rullani, E., (1997). Division of labour and the locus of inventive activity. *Journal of Management and Governance* 1(1): 123–140.

Bahemia, H., Squire, B., (2010). A contingent perspective of open innovation in new product development projects. *International Journal of Innovation Management* 14(4): 603–627.

Bakker, R., (2010). Taking stock of temporary organizational forms : A systematic review and research agenda. *International Journal of Management Reviews* 12(4): 466–486.

Battistella, C., Nonino F., (2012). Exploring the impact of motivations on the attraction of innovation roles in open innovation web-based platforms. *Production Planning and Control* 24(2–3) 226–245.

Bonesso, S., Comacchio, A., Pizzi, C., (2011). Technology sourcing decisions in exploratory projects. *Technovation* 31(10–11): 573–585.

Bosch-Sijtsema, P.M., Postma, T.J.B.M., (2009). Cooperative innovation projects: Capabilities and Governance Mechanisms. *Journal of Product Innovation Management* 26(1): 58–70.

Brusoni, S., Prencipe, A., Pavitt, K., (2001). Knowledge specialization, organization coupling, and the boundaries of the firm: Why do firms know more than they make? *Administrative Science Quarterly* 46(4): 597–625.

Brusoni, S., Prencipe, A., Salter, A., (1998). Mapping and measuring innovation in project-based firms, *CoPS Working Paper* No. 46, SPRU, University of Sussex.

Carson, S.J., Madhok, A., Wu, T., (2006). Uncertainty, opportunism, and governance: The effects of volatility and ambiguity on formal and relational contracting. *Academy of Management Journal* 49(5): 1058–1077.

Cassiman, B., Di Guardo, M.C., Valentini, G., (2010). Organizing links with science: Cooperate or contract? A project-level analysis. *Research Policy* 39(7): 882–892.

Cattani, G., Ferriani, S., Frederiksen, L., Täube, F., (2011). *Project-based organizing and strategic management.* Howard House, Wagon Lake, UK: Emerald Group Publishing.

Chakrabarti, A.K., Hauschildt, J., (1989). The division of labour in innovation management, *R&D management* 19(2): 161–171.

Chesbrough, H.W., (2003). *Open innovation: The new imperative for creating and profiting from technology.* Boston: Harvard Business School Press.

Chesbrough, H.W., (2006). *Open business models: How to thrive in the new innovation landscape.* Boston: Harvard Business School Press.

Chiaroni, D., Chiesa, V., Frattini, F., (2010). Unravelling the process from closed to open innovation: Evidence from mature, asset intensive industries. *R&D Management* 40(3): 222–245.

Chiaroni, D., Chiesa, V., Frattini, F., (2011). The open innovation journey: How firms dynamically implement the emerging innovation management paradigm. *Technovation* 31(1): 34–43.

Christiansen, J.K., Gasparin, M., Varnes, C.J., (2013). Improving design with open innovation a flexible management technology. *Research-Technology Management* 56(2): 36–44.

Clark, K.B., (1995). *The Product Development Challenge: Competing Through Speed, Quality, and Creativity,* Boston: Harvard Business School Press.

Cohen, W.M., Levinthal, D.A., (1990). Absorptive capacity: A new perspective on learning and innovation. Administrative Science Quarterly 35(1): 128–152.

Colombo, M.G., (2003). Alliance form: a test of the contractual and competence perspectives. *Strategic Management Journal* 24(12): 1209–1229.

Dittrich, K., Duysters, G., (2007). Networking as a means to strategy change: The case of open innovation in mobile telephony, *The Journal of Product Innovation Management* 24(6): 510–521.

Elmquist, M., Fredberg, T., Ollila, S., (2009). Exploring the field of open innovation. *European Journal of Innovation Management* 12(3): 326–345.

Foss, N.J., Husted, K., Michailova, S., (2010). Governing knowledge sharing in organizations: Levels of analysis, governance mechanisms, and research directions. *Journal of Management Studies* 47(3): 455–482.

Gardner, H.K., Gino, F., Staats, B.R., (2012). Dynamically integrating knowledge in teams: Transforming resources into performance. *Academy of Management Journal* 55(4): 998–1022.

Gassmann, O., Enkel, E., Chesbrough, H., (2010). The future of open innovation. *R&D Management* 40(3): 213–221.

Gassmann, O., Zeschky, M., (2008). Opening up the solution space: The role of analogical thinking for breakthrough product innovation. *Creativity and Innovation Management* 17(2): 97–106.

Gemünden H.G., Killen C., Kock A., (2013). Implementing and Informing Innovation Strategies through Project Portfolio Management. *Creativity and Innovation Management* 22(1): 103–104.

Goffin K., Koners, U., (2011). Tacit knowledge, lessons learnt, and new product development. *Journal of Product Innovation Management* 28(2): 300–318.

Grandori, A., Kogut, B., (2002). Dialogue on organization and knowledge. *Organization Science* 13(3): 224–231.

Grant, R.M., Baden-Fuller, C., (2004). A knowledge accessing theory of strategic alliances. *Journal of Management Studies* 41(1): 61–84.

Grönlund, J., Sjödin, D.R., Frishammar, J., (2010). Open Innovation and the Stage-Gate Process: A revised model for new product development. *California Management Review* 52(3): 106–131.

Gulati, R., Puranam, P., Thusman, M., (2012). Meta-organization design: rethinking design interorganizational and community contexts. *Strategic Management Journal* 33(6): 571–586.

Gulati, R, Kletter, D., (2005). Shrinking core-expanding periphery: the relational architecture of high performing organizations. *California Management Review* 47(1): 77–104.

Hagedoorn, J., (2002). Inter-firm R&D partnerships: An overview of major trends and patterns since 1960. *Research Policy* 31(4): 477–492.

Hagedoorn, J., Lorenz-Orlean, S., van Kranenburg, H., (2008). Inter-firm technology transfer: partnership-embedded licensing or standard licensing agreements? *Industrial and Corporate Change* 18(3): 529–550.

Hanisch, B., Wald, A., (2012). A bibliometric view on the use of contingency theory in project management research. *Project Management Journal* 43(3): 4–23.

Hobday, M., (2000). The project-based organization: An ideal form for managing complex products and systems? *Research Policy* 29(7–8): 871–893.

Howells, J., (2006). Intermediation and the role of intermediaries in innovation. *Research Policy* 35(5): 715–728.

Hsieh, K.N., Tidd, J., (2012). Open versus closed new service development: The influences of project novelty. *Technovation* 32(11): 600–608.

Huang, Y., Chung, H., Lin, C.H., (2009). R&D sourcing strategies: Determinants and consequences. *Technovation* 29(3): 155–169.

Huizing, E.K.R.E., (2011). Open innovation: State of the art and future perspectives. *Technovation* 31(1): 2–11.

Jurgens, U., (1999). Anticipating problems with manufacturing during product development process, in Comacchio A., Volpato G., Camuffo A., (eds) *Automation in automotive industries. Recent developments*, Berlin: Springer Verlag.

Knudsen, M.P., Mortensen, T.B., (2011). Some immediate-but negative-effects of openness on product development performance. *Technovation* 31(1): 54–64.

Kogut, B., Zander, U., (1992). Knowledge of the firm, combinative capabilities, and the replication of technology. *Organization Science* 3(3): 383–397.

Lakhani, K.R., Tushman, M.L., (2012). Open Innovation and Organizational Boundaries: The Impact of Task Decomposition and Knowledge Distribution on the Locus of Innovation, *HBS Working Paper* 12–057.

Langlois, R.N., Robertson, P.L., (1992). Networks and innovation in a modular system: Lessons from the microcomputer and stereo component industries. *Research Policy* 21(4): 297–313.

Laursen, K., Salter, A., (2006). Open for innovation: The role of openness in explaining innovation performance among U.K. manufacturing firms. *Strategic Management Journal* 27(2): 131–150.

Leiponen, A., Helfat, C.E., (2010). Innovation objectives, knowledge sources, and the benefits of breadth. *Strategic Management Journal* 31(2): 224–236.

Lenfle, S., (2008). Exploration and project management. *International Journal of Project Management* 26(5): 469–478.

Lewin, A.Y., Massini, S., Peeters, C., (2011). Microfoundations of internal and external absorptive capacity routines. *Organization Science* 22(1): 81–98.

MacCormack, A., Verganti, R., Jansiti, M., (2001). Developing products on "Internet Time": The anatomy of a flexible development process. *Management Science* 47(1): 133–150.

Mol, M.J., (2005). Does being R&D intensive still discourage outsourcing? Evidence from Dutch manufacturing. *Research Policy* 34(4): 571–582.

Mortara, L., Minshall, T., (2011). How do large multinational companies implement open innovation? *Technovation* 31(10–11): 586–597.

Nilsen, E.R., (2013). Organizing for learning and knowledge creation are we too afraid to kill it? Projects as a learning space. *International Journal of Managing Projects in Business* 6(2): 293–309.

Nooteboom, B., Haverbeke, W.V., Duysters, G., Gilsing, V., van den Oord, A., (2007). Optimal cognitive distance and absorptive capacity. *Research Policy* 36(7): 1016–1034.

Oxley, J., (1997). Appropriability hazards and governance in strategic alliances: A transaction cost approach. *Journal of Law, Economics, & Organization* 13(2): 387–409.

Powell, W.W., Koput, K.W., Smith-Doerr, L., (1996). Interorganizational collaboration and the locus of innovation: Networks of learning in biotechnology. *Administrative Science Quarterly* 41(1): 116–145.

Powell, W.W., White, D.R., Koput, K.W., Owen-Smith, J., (2004). Network dynamics and field evolution: The growth of inter-organizational collaboration in the life sciences. *American Journal of Sociology* 110(4): 1132–1205.

Prencipe, A., (2000). Breadth and depth of technological capabilities in CoPS: The case of the aircraft engine control system. *Research Policy* 29 (7–8): 895–911.

Prencipe, A., (2003). Corporate Strategy and Systems Integration Capabilities Managing Networks in Complex Systems Industries, in Prencipe, A., Davies, A., Hobday, M. (eds). *The Business of Systems Integration.* (pp. 114–132) Oxford: Oxford University Press.

Rass, M., Dumbach, M., Danzinger, F., Bullinger, A.C., Moeslein, K.M., (2013). Open innovation and firm performance: The mediating role of social capital. *Creativity and Innovation Management* 22(2): 177–194.

Reich, B.H., Gemino, A., Sauer, C., (2012). Knowledge management and project-based knowledge in it projects: A model and preliminary empirical results. *International Journal of Project Management* 30(6): 663–674.

Roijakkers, N., Hagedoorn, J., (2006). Inter-firm R&D partnering in pharmaceutical biotechnology since 1975: Trends, patterns, and networks. *Research Policy* 35(3): 431–446.

Rothaermel, F.T., Deeds, D.L., (2004). Exploration and exploitation alliances in biotechnology: A system of new product development. *Strategic Management Journal* 25(3): 201–221.

Salge, T.O., Farchi, T., Barrett, M.I., Dopson, S., (2013). When does search openness really matter? A contingency study of health-care innovation projects. *The Journal of Product Innovation Management* 30(4): 659–676.

Söderlund, J., Vaagaasar, A.L., Andersen, E.S., (2008). Relating, reflecting and routinizing: Developing project competence in cooperation with others. *International Journal of Project Management* 26(5): 517–526.

Takeishi, A., Fujimoto, T., (2003). Modularization in the car industry: interlinked multiple hierarchies of product, production, and supplier systems, in Prencipe, A., Davies, A. Hobday, M. (eds). *The Business of Systems Integration.* (pp 254–278). Oxford: Oxford University Press.

Tidd, J., Bessant, J., Pavitt, K., (2005). *Managing innovation integrating technological, market and organizational change.* Chichester: John Wiley & Sons Ltd.

Tranekjer, T.L., Søndergaard, H.A., (2013). Sources of innovation, their combinations and strengths—benefits at the NPD project level. *International Journal of Technology Management* 61(3–4): 205–236.

Trombini, G., Comacchio, A., (2012). Cooperative Markets for Ideas: When does Technology Licensing Combine with R&D Partnerships?, *DRUID Conference,* Copenhagen Business school, Copenhagen; Denmark, 19–21 June 2012.

Tsai, K-H., (2009). Collaborative networks and product innovation performance: Toward a contingency perspective.*Research Policy* 38(5): 765–778.

van de Vrande, V., de Jong J.P.J., Vanhaverbeke, W., de Rochemon, M., (2009). Open innovation in SMEs: Trends, motives and management challenges. *Technovation* 29(6–7): 423–437.

Von Hippel, E., (2005). *Democratizing Innovation: The Evolving Phenomenon of User Innovation.* Cambridge, Massachusetts: The MIT Press.

von Zedtwitz, M., Gassmann, O., (2002). Market versus technology drive in R&D internationalization: four different patterns of managing research and development. *Research Policy* 31(4): 569–588.

West, J., Salter, A., Vanhaverbeke, W., Chesbrough, H., (2014). Open innovation: The next decade. *Research Policy* 43(5): 805–811.

Wheelwright, S.C., Clark, K.B., (1995). Creating Project Plans to Focus Product Development, in Clark, K.B., Wheelwright, S.C., (eds) *The Product Development Challenge: Competing Through Speed, Quality, and Creativity.* (pp. 187–208). Boston: Harvard Business School Press.

Zhang, J., Baden-Fuller, C., Mangematin, V., (2007). Technological knowledge base, R&D organization structure and alliance formation: Evidence from the biopharmaceutical industry. *Research Policy* 36(4): 515–528.

Chapter 2
Exploring the Knowledge Space Through Project-Based Sourcing

Sara Bonesso, Anna Comacchio and Claudio Pizzi

Abstract Only recently has open innovation research emphasized the relevance of adopting a project-level contingency approach for explaining inbound sourcing choices. Our research aims to add to this issue by providing new insights on the knowledge-based determinants of sourcing decisions at the project level of analysis. We maintain that a new product development (NPD) project can be conceived as a strategic means not only to explore the knowledge space for the identification of high-value solutions, but also to search the sources that enable the firm to develop the specific knowledge features. We suggest that the knowledge space explored by an NPD project is grounded on the main elements of an industrial innovation system and that it is characterized by two key dimensions, namely knowledge novelty, the knowledge space of the performance features of a product that meet new customer needs, and knowledge breadth, the knowledge space of technological domains to draw on for solving product-related problems. Our research is implemented on a sample of NPD projects carried out by a group of leading Italian firms, operating in the machine tool industry. Findings show that in companies which define sourcing on a project-by-project basis, projects that explore at the frontier of either novel product features or heterogeneous technological domains, spur firms to rely on external sources and to choose R&D development agreements as the governance form to involve partners. Moreover, a high degree of knowledge novelty induces firms to search cognitive distant partners instead of similar ones. Proposing a project-based approach to strategi-

S. Bonesso (✉) · A. Comacchio
Department of Management, Ca' Foscari University of Venezia,
San Giobbe Cannaregio, 873, 30121 Venezia, Italy
e-mail: bonesso@unive.it

A. Comacchio
e-mail: comacchio@unive.it

C. Pizzi
Department of Economics, Ca' Foscari University of Venezia,
San Giobbe Cannaregio, 873, 30121 Venezia, Italy
e-mail: pizzic@unive.it

S. Bonesso et al. (eds.), *Project-Based Knowledge in Organizing Open Innovation*,
DOI: 10.1007/978-1-4471-6509-5_2, © Springer-Verlag London 2014

cally organize inbound sourcing, the chapter provides evidence on the concept of a company sourcing strategy as a portfolio of decisions across projects.

2.1 Introduction

Open innovation research emphasizes the benefits and the drawbacks a firm may encounter in combining externally generated knowledge with that accumulated inside (Garriga et al. 2013; Knudsen and Mortensen 2011), but scant attention has been paid to the question of how firms make the decision to open up their innovation process (Hsieh and Tidd 2012). Specifically, this stream of the literature has not devoted enough attention to the drivers that explain the choice to rely on external sources and the related decisions concerning partner selection and the appropriate modes of governance of inter-organization knowledge production and acquisition. In this regard, the extant empirical evidence has considered the endowment of a company as the explanation for sourcing decisions, investigating how past accumulation of knowledge can be a premise that allows a firm to recognize and absorb external knowledge (Nooteboom et al. 2007; Tsai 2009; Zhang and Baden–Fuller 2010).

While the centrality of a project as a means of internal knowledge production is widely recognized by the literature on innovation and project management, this level of analysis is neglected in studies on how a firm engages external sources in the exploration and production of new knowledge, for new product development (NPD).

Only recently have studies emphasized the relevance of adopting a project-level contingency approach for explaining inbound open innovation (Bahemia and Squire 2010; Bonesso et al. 2011; Salge et al. 2013; Tranekjer and Søndergaard 2013). Our research aims to add to this issue by providing new insights into the knowledge-based determinants of inbound sourcing decisions at the project level of analysis. We maintain that the NPD project can be conceived as a strategic means not only to explore the knowledge space for the identification of high-value solutions to create new products (Macher 2006; Terwiesch and Xu 2008), but also to search the sources that enable the firm to develop the specific knowledge features required by an NPD project.

The remainder of this chapter is organized as follows. The next section introduces the project-based approach in studying inbound open innovation. Drawing on the sectoral innovation system framework, the subsequent section describes the attributes of the knowledge space that an NPD project can explore, namely knowledge novelty and knowledge breadth. Successively, we formulate the theoretical arguments underpinning the hypotheses on the impact of each knowledge attribute on the three main inbound sourcing choices, namely the decision: (1) to tap external rather than exclusively internal knowledge sources; (2) to co-develop the NPD project with external partners; (3) to rely on cognitive distant sources rather than on similar ones. Next, we describe the research setting, data sources, the variables included in the study, and the estimation methods. After presenting the most relevant results, in the final section, we discuss the findings and draw conclusions, proposing a project-based approach to strategically organize inbound sourcing.

2.2 Inbound Open Innovation at Project Level: A Knowledge-Based Perspective

Opening up the firm innovation process through inbound activities stimulates the generation of new knowledge by developing in-house core competencies and combining a diverse pool of complementary sources. This may lead to increased product portfolio diversity, better matching of the firm's offer and consumer needs, and consequently higher innovation performance (Laursen and Salter 2006; Parida et al. 2012; van de Vrande et al. 2009). The strategic organization of how firms get access to external new knowledge and integrate it internally represents a central topic in the recent debate on open innovation research (Gassmann et al. 2010). Sourcing decisions, related to the appropriate forms of governance, as well as partner selection have been analysed adopting primarily a transaction costs approach. This approach defines on the one hand advantages, in terms of R&D costs and risk sharing, and on the other hand, barriers, related to partner selection, and coordination, as well as risks of knowledge leakages and imitation (Becerra et al. 2008; Huang et al. 2009; Mol 2005; Mowery et al. 1998; Robertson and Gatignon 1998). While the debate on the impact of economic factors has advanced the understanding on external sourcing determinants to efficiently exploit partners and safeguard from opportunistic behaviours, it underestimates the role played by knowledge and its attributes in sourcing decisions. The main criticism is that this approach does not consider the strategic opportunity of knowledge creation through partnership (Zajac and Olsen 1993).

Through the knowledge-based perspective lens, a firm opening up its organizational boundaries searches for complementary external knowledge to create new products (Katila and Ahuja 2002) by strategically designing the external network of knowledge sources with which it could create new value (Zajac and Olsen 1993).

From this perspective, the innovation process can be represented as a knowledge search activated either at firm or at project level. The knowledge space, which a firm aims to explore by its search for novel knowledge and partners, is a sectoral or inter-sectoral competitive landscape (Danneels and Kleinschmidt 2001). Indeed, the sectoral innovation system sets the innovation opportunities and constraints (Malerba 2002, 2005) along the two main axes of market needs and technological solutions (Brunswicker and Hutschek 2010). Drawing on prior studies, we suggest that the main attributes of the knowledge space explored by a firm are related to these two coordinates (Bonesso et al. 2011; Nickerson and Zenger 2004; Terwiesch and Xu 2008) and consequently they affect how a firm strategically searches for new knowledge and organizes its network of external partners.

The open innovation literature has contributed to advancing the understanding of how a firm combines internal and external knowledge to create new value through innovation processes spanning organizational boundaries; however, less attention has been paid to the drivers that impact on the firm's decision to open up the innovation process. Moreover, research has remained silent on the knowledge attributes of new products that a firm aims to develop through sourcing activities and on how these attributes might impact on the decisions to cross organizational boundaries. Indeed, this stream of the literature mainly focuses on the innovation strategy of the firm (closed versus open approach), explaining the implementation process of business models and the consequent organizational solutions (Chesbrough 2006; Chiaroni et al. 2010; Mortara and Minshall 2011).

A few empirical studies have recently advanced the understanding of the knowledge attributes as inbound sourcing determinants (van de Vrande et al. 2009; Zhang and Baden–Fuller 2010). While these studies have provided new insights, they mainly focus on the characteristics of the knowledge base of the firm, rather than on the project level of analysis. As claimed by recent research "contingency studies on open innovation are hence needed especially at the project level" (Salge et al. 2013, p. 660), for three main reasons.

First, the NPD project represents the locus where knowledge exploration and production is primarily carried out (Lenfle 2008).

Second, we argue that a study of inbound open innovation at project level is sound due to the fact that the central inbound choices at the NPD project level concern whether, how and where to tap specialized external sources coherently with the knowledge attributes explored by the project. In the case of non-incremental NPD projects, a firm pursues new objectives by developing novel components and product architecture (Henderson and Clark 1990). Thus, it may be spurred to search for new knowledge not only beyond its organizational boundaries, but also by adopting sourcing decisions independently from those made in the past. This means it might define project-by-project the features of a product and the range of knowledge sources it wants to draw on (Bonesso et al. 2011; Knudsen and Mortensen 2011; Tranekjer and Søndergaard 2013).

Concerning the third reason, we suggest that in line with studies on the project portfolio strategy of a firm (Knudsen and Mortensen 2011), the adoption of the project level of analysis to study inbound sourcing not only enhances the understanding of the sourcing decision of any single project, but also provides primary explorative evidence on the concept of a company sourcing strategy as a portfolio of decisions across projects.

Our study, bridging the literatures on open innovation, strategic project management and the knowledge-based view, aims to investigate the effects of the knowledge attributes a firm wants to develop project-by-project on inbound open innovation. In particular, we aim to advance the open innovation research, which is mainly focused on the firm level of analysis, adopting a project-based approach in studying the determinants of external sourcing. On the other hand, we want to add to the project management literature, considering the project not only a means to manage an NPD process efficiently and effectively but also as a vehicle to make sourcing portfolio decisions. Finally, we want to extend the knowledge-based studies by offering an conceptualization and operationalization of the knowledge attributes a project aims to generate.

2.3 Defining the Knowledge Space at Micro-Level: Knowledge Attributes Explored by NPD Projects

As claimed by Lenfle (2008, p. 471) "the result of the project is then no longer simply a product" but the opportunity to learn new knowledge that can significantly foster the capacity to be innovative (Söderlund et al. 2008). Firms generate new knowledge by selecting a problem to solve and starting an exploration process of valuable and innovative knowledge combinations (Macher 2006). When a firm wants to solve product-related problems it might engage in a search process by launching a new project. Therefore, the attributes of the knowledge space explored are not related to the stock of knowledge accumulated by the firm, but are those that characterize the new knowledge the firm aims to develop by the problem-solving process activated in each NPD project. These attributes can be conceived as the coordinates of the knowledge space (Terwiesch and Xu 2008) within which a project "engages in a process of search for high-value solutions" (Macher 2006, p. 827).

As suggested by prior studies, the knowledge space explored by an NPD project is grounded on the main elements of an industrial innovation system (Malerba 2002, 2005), which can be conceived as the landscape (Nickerson and Zenger 2004) within which firms aim to discover new knowledge combinations through the launch of NPD projects. We suggest that this landscape is structured around two key knowledge dimensions, namely knowledge novelty, the knowledge space of the performance features of a product that meet new customer needs, and knowledge breadth, the knowledge space of technological domains to draw on for problem solving (Bonesso et al. 2011; Danneels and Kleinschmidt 2001).

Knowledge novelty can be defined as a knowledge attribute which provides superior product functionalities for customers and thus improvements in performance features (Amara et al. 2008). Exploring knowledge novelty implies a process of product concept shift (Seidel 2007) or ideation (Dahl and Moreau 2002) that helps to depart from the existing industry offering. Indeed, knowledge novelty is a matter of degree (Freel and de Jong 2009), since if the project explores the space of customer problems and needs in order to develop a novel concept and new functionalities not available in the industry, this means that the project presents a high degree of knowledge novelty. On the other hand, a project presents a lower extent of novelty if new features are introduced into a firm's portfolio for the first time, but are already available on the market. In the latter case, knowledge novelty is not explored at its frontier. High-novelty projects develop original concepts and features by addressing problems not already solved by competitors and in so doing they satisfy emergent needs. For this reason they are usually positively associated with higher returns (Marsili and Salter 2005). Departing from the existing industrial solutions entails a stronger effort in the exploration of the solution space, in terms of time and resources devoted to scouting, understanding, evaluating and exploiting market opportunities for new functions which are not yet available in the existing products of the same industry.

Independently of the degree of novelty, new functionalities imply a problem-solving process: the expected features are carried out by elements whose operating principles are based on a scientific and technological domain (Brusoni and Prencipe 2006). Since a technological domain is "a group of technologies that solve primary problems" (George et al. 2008, p. 1449), the knowledge breadth of a project can be conceived in terms of the degree of diversity or heterogeneity among technological domains a project draws on to solve different primary problems (Wang and von Tunzelmann 2000). The dynamic transformation of several sectors towards technology fusion (e.g. mechatronics, biopharmaceuticals, optoelectronics) (Kodama 1992) implies a convergence and an integration of previously separated knowledge and technologies (Malerba 2005), which increases the heterogeneity of primary problems and the domains a firm may draw on through NPD projects. Empirical evidence confirms the blurring of boundaries between technological disciplines (Choi and Valikangas 2001) in high–tech as well as in low–medium-tech sectors (Bröring and Leker 2007; Freddi 2009; Wengel and Shapira 1994). This implies that sector-specific technological domains (for instance, chemistry in the pharmaceutical industry, mechanics in the equipment industry) are combining with diverse technological and scientific disciplines which have progressively been added to the search space that a firm can investigate through NPD projects (biology in the pharmaceutical industry, electronics and software in the equipment industry) (Gambardella and Torrisi 1998; Quintana–Garca and Benavides–Velasco 2008). Therefore, firms facing the challenge of technological fusion may need to master through an NPD project a wider range of disciplines than in the past. We claim that the integration of heterogeneous disciplines in an NPD project increases the extent to which the knowledge investigated by that project can be conceived as broad. On the other hand, knowledge breadth can be conceived as narrow, when the NPD project explores the consolidated industrial scientific and technological knowledge. Any different additional domain included in the search space of an NPD project expands the horizon for opportunities to scan for a new knowledge combination, but it also enhances the difficulties in understanding interdependencies among a wider range of interrelated problem settings.

In the next section, we present our theoretical arguments, suggesting that inbound sourcing decisions are contingent to the NPD features, and specifically to the degree of knowledge novelty and breadth explored by an NPD project.

2.4 Knowledge Attributes of an NPD Project and Inbound Sourcing

Defining the composition of the sourcing portfolio has become an important part of a firm's overall strategy (van de Vrande 2013, p. 610). Although research demonstrates the benefit of having a diversity of sourcing portfolios depending on different circumstances, such as the degree of similarity between the firms and the external partners, the analysis of sourcing composition has not been considered in relation to the project portfolio characteristics.

Our research extends the literature on the contingent factors that influence the decision-making process of inbound sourcing, investigating the impact of the degree of knowledge novelty and breadth explored by the NPD projects on three main choices:

- whether to rely on external partners instead of relying exclusively on in-house sources;
- how to get access to the knowledge (modes of governance to implement);
- where to source the knowledge (distant versus similar sources).

2.4.1 Internal Versus External Sourcing

When a firm engages in an NPD project, exploring at the frontier of knowledge novelty (searching for radically new product concepts and functionalities to satisfy new needs) or knowledge breadth (searching for technological solutions in a heterogeneous technological and scientific space), it may be induced to open up its innovation process for valuable interactions with competent external sources.

The generation of original product features requires the adoption of a divergent way of thinking which implies the development of a wide range of non-conventional ideas (Colarelli O'Connor 1998). The exclusive reliance on internal sources may spur towards a convergent way of thinking; instead, interaction with external partners may not only enlarge the search space in terms of number of ideas providing room for inspiration (Freel and de Jong 2009), but also encourage divergent thinking through the departure from characteristics in the specific sector (high knowledge novelty). Studies have highlighted the relevance of the use of analogies, as a means of creative thinking for problem solving, to convey novelty in NPD projects (Dahl and Moreau, 2002; Gassmann and Zeschky 2008; Kalogerakis et al. 2010). The term "analogy" refers to the successful identification of similarities (superficial or structural) between a source and a target domain (Gentner 1983). Interaction with external sources might enhance problem-solving effectiveness and efficiency in terms of identification of far analogous solutions. Indeed, external partners may act as brokers, on the one hand making non-obvious connections between different categories of products which share some similarities, and on the other enabling the combination of functionalities not previously introduced into the projects of the firm-target's industrial context (Hagardon and Sutton 1997).

Moreover, external sourcing may transfer to producers the advanced experiences of innovative "lead users", who aim to solve their own ahead-of-market needs. In this regard, it has been demonstrated that in the process equipment or software sector, innovations transferred from users "tended to be those of stronger and more general interest to users, and thus of more value to producers as commercial products" (de Jong and von Hippel 2009, p. 1181). Therefore, this external technology source reduces the level of uncertainty of market acceptance of newness. Besides this advantage, sourcing user innovation in high-novelty projects enables reductions in engineering-related costs and risks due to the fact that the lead user has already carried out some preliminary prototyping tests (de Jong and von Hippel 2009). This is the case of Business-to-Business (B2B) producers whose lead users have the capabilities to anticipate and solve their own ahead-of-market needs (Robertson et al. 2003). Thus, we may expect that:

Hypothesis 1 The higher the knowledge novelty in an NPD project, the more likely the external sourcing.

When the degree of an NPD project's technological heterogeneity is high, the risks and costs of a search process in specific technological and scientific domains are better managed when they are partitioned among specialized partners. Time-to-market of NPD projects with high knowledge breadth may be decreased by external sourcing choices since knowledge suppliers match solutions and problems faster due to their experience curve. External sourcing also impacts positively on production costs because specialized suppliers in different disciplines may exploit economies of scale, since they can spread their investments over a larger base of development activities (Macher 2006). Moreover, the incentives to overcome barriers against external sources are even higher when the pace of change in non-core technology fields rises and firms need to keep up at the edge of all these fields (Mol 2005) without bearing the risk of the exploration process across different scientific frontiers. Thus, in the fast-changing technology landscape (Fleming and Sorenson 2003), it could be more convenient to adopt a flexible approach to sourcing by exploiting a partner's capacity to be at the frontier of a specific technological domain, avoiding at same time the high investments and sunk costs of in-house R&D.

Moreover, when an NPD project is characterized by diverse primary problems which can be solved through a search process in heterogeneous domains, a number of potential interdependencies arise among solutions offered by each single technological field. Problem-solvers face relevant constraints in structuring a problem which spans over multiple knowledge sets, due to the low understanding of the map of possible interdependencies (Macher 2006). Therefore, firms may prefer to focus their limited efforts and resources, on the one hand, on the search activity in the consolidated scientific and technological knowledge of the sector and, on the other hand, on the management of knowledge integration problems, while relying on specialized partners for solution-seeking within each additional domain. Thus:

Hypothesis 2 The higher the knowledge breadth in an NPD project, the more the external sourcing.

2.4.2 How to Source? Inbound Sourcing Through R&D Development Agreements

The exploration at the frontier of the knowledge space (high knowledge novelty or high knowledge breadth) may entail a significant cognitive endeavour that can jeopardize the recognition and the implementation of valuable solutions to innovation problems. Open innovation literature shows that non-equity-based collaborative relationships favour the process of exploration of market and technology opportunities and seem to offer flexibility, speed, and innovation (Dittrich and Duysters 2007; Laursen et al. 2010; van de Vrande 2013). Therefore, we maintain that firms exploring at the frontiers of the knowledge space through NPD projects may reduce these cognitive constraints and increase learning opportunities through R&D development agreements with external partners.

Recent studies show that novel ideas in terms of new product functionalities and performance features emerge from the original combination of pieces of knowledge across industries through far analogies (Brunswicker and Hutschek 2010). The identification of non-obvious analogies in the market offering of different industries brings higher customer benefits (high knowledge novelty) than those based on near analogies (Kalogerakis et al. 2010), but they "are more difficult to identify and require more cognitive effort" (Gassmann and Zeschky 2008, p. 98). As discussed by prior research, the successful identification of far analogies and their subsequent translation may require an interactive and mutual learning process between the seeker-source target and the solver-source domain (Enkell and Gassmann 2010). As suggested by Nooteboom et al. (2007, p. 1017) "When people with different knowledge and perspectives interact, they stimulate and help each other to stretch their knowledge for the purpose of bridging and connecting diverse knowledge". Therefore, the relationship between the firm and the external sources involved in the NPD cannot be treated purely as a transaction if the project aims to depart from the existing industry offering, but requires forms of co-development that makes it possible to better detect similarities (in terms of product features and functionalities) between unrelated domains and effectively transfer the contents to the target-firm's product features.

Moreover, firms exchanging knowledge with a partner in an early stage and at the frontier of the knowledge domain, might face high degree of ambiguity and it might be difficult to communicate and share sticky and contextual knowledge. Thus, they need to rely on appropriate coordination mechanisms and incentives to access the partner's skills and optimally internalize the exchanged technology (Trombini and Comacchio 2012). Hence:

Hypothesis 3 The higher the knowledge novelty in an NPD project, the more likely the R&D development agreement.

An important driver in implementing R&D development agreements is related to the complexity of the problem to be solved, which is higher when heterogeneous domains need to be explored (high knowledge breadth) and the understanding of the interdependencies among them is low (Macher 2006; Simon 1962). In order to reduce uncertainty and increase the understanding of the relationships between different technological domains, the engagement of external partners in the NPD project may be beneficial. R&D agreements imply frequent contacts that stimulate mutual understanding as well as the development of a common language and a communication code that can facilitate joint problem-solving and reduce the time and the cost related to the integration of different technological domains (Hsieh and Tidd 2012).

Moreover, firms may be willing to keep abreast by interacting with partners more expert in other technological fields. The learning process that can be activated by an R&D co-development might help to accumulate in-house basic knowledge in a diverse discipline, increasing a firm's familiarity with it, useful for future search processes. Therefore:

Hypothesis 4 The higher the knowledge breadth in an NPD project, the more the R&D development agreement.

2.4.3 Where to Source? Similar Versus Distant Partners

Not only can firms define project-by-project whether and how to involve external partners in the NPD process but also where to search for the potential solvers of product-related problems concerning both market needs and technological solutions.

Studies on partner identification and selection highlight the importance of similarities among partners in terms of shared goals and convergent interests as well as norms of behaviour that facilitate coordination, reduce risks of opportunism thanks to the development of close trust-based relationships and accelerate the learning process (Cummings and Holmberg 2012; Rothaermel and Boeker 2008).

Despite the positive effects of cognitive alignment or proximity among partners, open innovation literature positively evaluates a moderated distance among firms and provides empirical evidence on its inverted U-shaped effect on innovation performance (Nooteboom et al. 2007). Moreover, research on geographical clusters and social capital has pointed out the negative side of a high level of cognitive proximity such as lock-in effects and redundant relationships that prevent new knowledge creation (Boschma 2005; Burt 2005; McEvily and Zaheer 1999). Therefore, we maintain that when an NPD project explores at the frontier of the knowledge space, a firm may benefit from cognitive distant partners in terms of opportunities for discovering original product features and functionalities which depart from its sector. These learning advantages can counterbalance the costs of overcoming the barriers related to the access of physically and culturally distant sources (Al–Laham and Amburgey 2011). A first benefit that cognitive distance yields is related to the access to different customers' systems of meanings and interpretation that help to identify and better define ahead-of-market needs and identify solutions to translate into the firm's product offering. Second, diversity between the target problem and the source domain may favour the process of detecting non-obvious analogies (Kalogerakis et al. 2010), whereas if source and target share the same conceptual domain they will lead to incremental innovation (Gassmann and Zeschky 2008). Therefore:

Hypothesis 5 The higher the knowledge novelty in an NPD project, the more likely distant partners are involved.

The literature on regional innovation systems demonstrates how geographical areas present a specific degree of expertise in technological and scientific disciplines related to a specific sector (Malerba 2004). This localized learning process provides a firm with the opportunity to interact with partners specialized in different technological fields all related to the same industrial cluster, with a high alignment in terms of shared goals and cultural norms. When a firm engages in an NPD project that requires a problem-solving process in additional technological domains (high knowledge breadth), the cognitive proximity between the firm and the sources may be beneficial for tapping the specialized language and mindsets of a specific technological domain. We argue that the higher degree of diversity of the technological domains explored by the NPD project may prevent the firm from searching physically and culturally distant partners in order to avoid adding further complexity in its exploration process. Indeed, knowledge creation and production may require not only the use of codified solutions but also of inductive activities of testing, experimentation, simulation and practical work (Asheim and Coenen 2005). Especially in the case of fusion among previously separated technological domains, technical solutions are often the result of experience gained through learning by doing and interacting. The cognitive proximity between the firm and the source specialized in the additional knowledge domain may enable an interactive and trust-based learning that favours the understanding of the interdependencies among disciplines and the management of integration problems. Thus, we may expect that:

Hypothesis 6 The higher the knowledge breadth in an NPD project, the more likely similar partners are involved.

2.5 Methods

2.5.1 Research Setting

The setting of our research is the machine tool industry, which is a long-established sector in the most advanced economies and still plays a pivotal role in Europe (Freddi 2009; Wengel and Shapira 2004). Specifically, we carried out a survey on a sample of NPD projects undertaken from 2002 to 2006 by seven leading medium Italian firms operating worldwide.

The machine tool industry represents an ideal context in which to investigate the impact of knowledge attributes on inbound sourcing at project level for three reasons. First, studies on industrial innovation systems confirm that firms in this sector are progressively opening up their innovation processes through collaboration with a variety of external partners (Wengel and Shapira 2004). Second, research activities and learning processes in the machine tool firms are typically performed on a project basis (project duration usually ranges from six months to over 1 year), thus the knowledge attributes of a project are salient. Finally, the two knowledge attributes of the projects, novelty and breadth, modeled as explanatory factors of inbound sourcing are particularly relevant in this industry, in which both demand requirements and heterogeneity of technological domains are increasingly compelling.

Concerning knowledge novelty, in this highly competitive B2B environment the key players nowadays are those firms able to innovate at the front-end, meeting emergent market demand instead of merely adopting an efficiency-based approach.

Empirical studies support this argument, highlighting that in this industry the rate of introduction of new products is high in comparison with that in other long-established sectors (MacPherson and Kalafsky 2003). This could be motivated by the fact that machine tools are capital goods (e.g. lathes, punching machines, press brakes, machining centres) central to almost all durable products. The literature places machine tool firms within the "enabling sectors" (Robertson et al. 2003) or the "specialized suppliers" (Pavitt 1984), namely suppliers of pervasive technologies (Brusoni and Sgalari 2006) that have a large influence on the manufacturing performances of other industries. The innovation process aims to increase the value of these capital goods for the users, especially for highly innovative clients such as the automotive, aeronautical, aircraft, aerospace and electricity supply sectors. Moreover, a notable characteristic of these products is their high durability, which would imply that a customer takes many years to place a new order. To increase the rate of substitution, companies are spurred to introduce significant advancements in the market in terms of the functionality of their machines.

From the point of view of knowledge breadth, this sector, since the introduction of computer-controlled devices in the 1980s, has been facing a technological shift from a dominant paradigm to a reconfiguration of the technical knowledge embodied in the product (Chen 2009; Sandven et al. 2001). According to Kodama (1992), machine tools are a typical example of a mechatronic product,[1] which is characterized by progressive integration of the traditional technological field, mechanics, with two different technological disciplines, namely electronics and software engineering (Freddi 2009; Wengel and Shapira 2004).

2.5.2 Data Collection

We obtained the list of the machine tool firms operating in the North East of Italy from the Association of Italian manufacturers of machine tools, robots, automation and ancillary products (numerical control systems, tools, components and accessories) UCIMU. According to the UCIMU Annual Report, in 2004 the Italian machine tool industry comprised 415 firms and employed 28,120 people; 15 % of total firms were located in the North East (Ucimu 2006). Initial contacts were made by e-mail and afterwards each firm's representative was called in order to present the aims of the study. Fourteen firms agreed to participate in the research.

Once consent had been obtained, we interviewed by phone the person responsible for the innovation activity of the firm, namely the R&D manager or the Engineering manager, in order to identify and assess the type of projects that had been started since 2002. Seven firms indicated that they had introduced only minor incremental

[1] A mechatronic machine/component was defined as "a mechanical element controlled by an electronic application that is integrated into it. Control means that the machine/component has the ability to change performance according to a change in external conditions. It is the high level of integration between the different technologies (mechanics, electronics and informatics) that distinguishes a mechatronic device from a mechanical, electronic or informatic one" (Freddi 2009, p. 552).

changes to their products during the period under examination. Due to the fact that our research focuses on projects which aim to develop knowledge which departs to some extent from that already embodied in the previous machines, these firms were discarded.

A total of 86 NPD projects, developed from 2002 to 2006, were obtained from the remaining seven medium-sized firms.

The dataset was constructed through several visits on site and phone contacts, drawing on multiple data sources.

A structured questionnaire with closed-ended questions was administered in order to collect data on the characteristics of the company, its R&D/Engineering department, and its products. Data were gathered from different respondents (the owner or the top management and the functional managers) according to the information required.

Data at the project level of analysis were collected through in-depth semi-structured interviews administered face-to-face. On each research site, the R&D/Engineering manager provided us with a list of all the projects the company had carried out since 2002, and which fitted the aims of our study, namely projects that did not introduce merely incremental changes, such as restyling of current product lines (Smith and Tushman 2005). All the projects identified by the respondents, which regarded a machine as a whole or as a set of components, can be considered successful from the market performance point of view. This can be explained by the fact that in this industry firms decide to invest in projects beyond the first stages when there is a preliminary sale agreement signed by a client, in consideration of the high economic value of this type of industrial equipment and the related investment required in the detailed design stage. We did not include in our analysis cases of project failure since the high costs of these machines led the project team to devote considerable efforts towards the detection of potential failures during the preliminary stages. Therefore, possible technical problems that may affect market performance of a new product are identified and resolved before sourcing decisions are made. The last column of Table 2.1 reports the number of projects by company.

We interviewed at least two knowledgeable informants per firm, all senior technicians, namely the engineering or the R&D manager and project leaders. The respondents were asked to describe in detail the content of each NPD project started between 2002 and 2006. Some examples include new technological principles (laser and plasma in cutting processes), materials (ecological and energy-saving treatments of natural resources), architectures or components (morphology that increases general performance, more precise and productive bending systems which integrate sophisticated electronic control devices). Afterwards, we collected fine-grained data on the two knowledge attributes under analysis and on the sourcing choices made for each project. The respondents were asked to describe in detail the sources that each NPD project drew on. The presence of multiple respondents allowed us to discuss potential disagreements (Miller et al. 1997). To limit common method variance problems, we collected the data on the dependent and independent variables at different times (Podsakoff and Organ 1986). This also gave the respondents time to search their memory and consult the necessary technical documentation to answer the questions on the project dimensions under investigation.

Finally, we drew on secondary data: (1) each firm's archive of product catalogues from the period under analysis which embodied the technical content developed by the projects; (2) articles from specialized magazines which reported the description of the firms' products; (3) the web site information of the leading international trade fairs where the companies presented their machines, and (4) discussion with external experts on the project description provided by the respondents. The use of multiple sources of information allowed a process of data triangulation (Sonali and Corley 2006), thus reducing potential bias deriving from an individual's memory failure and protection mechanisms and ensuring the internal validity of the measures regarding project novelty and breadth.

2.5.3 Variables

2.5.3.1 Dependent Variables

Internal versus external sourcing Drawing on the classifications traditionally proposed in the literature, we identified two categories of sourcing choices that may be implemented in each project: internal and external (Cassiman and Veugelers 2006; Veugelers and Cassiman 1999). Internal sourcing includes the firm's own R&D and technological transfer and assistance from parent or associate companies. External sourcing encompasses a wide range of modes: arm's length arrangements, which refer to unilateral knowledge flows (licensing agreements and purchasing from supply chain actors), intermediate mechanisms between market and hierarchy, namely R&D cooperation with other firms, and the acquisition of other companies prompted by the requirements of an ongoing project (Narula and Hagedoorn 1999; van de Vrande et al. 2009; Zhao et al. 2005). We constructed a binary variable that takes the value 1 when the project involved external sources. Whereas when the project was developed relying exclusively on internal sources, such as the firm's own R&D department and transfer from parent firms, the variable takes the value 0.

R&D development agreements Sourcing can be achieved through the use of different modes of governance with diverse implications in terms of opportunities for inter-firm learning (Narula and Hagedoorn 1999; van de Vrande 2013). Our respondents were asked to describe the forms of governance each NPD project used for its development. In our sample we distinguished between projects that do not use modes of governance that enable the activation of an inter-firm learning process (market transaction and in-house development) and projects that involved the partners in forms of agreements characterized by a high level of inter-firm interaction. We constructed a binary variable that takes the value 1 when the project involved for its development an R&D cooperation based on contractual agreements with external sources (suppliers, clients, universities, consultants, etc.), otherwise the variable takes the value 0.

Partner distance We measured the difference among firms in terms of cultural and physical distance between the firm and the sources involved in the NPD project (Boschma 2005; Teixeira et al. 2008) as a proxy for cognitive distance. The respondents were asked to indicate the geographical localization of the sources involved in the development of each NPD project, identifying in each project the nationality of the sources. The partner distance has been measured on a 3-point scale, where 0 means that the project does not draw on sources beyond the firm's boundaries, thus there is no cognitive distance; 1 means that the project involves only sources geographically located in the same country (Italy), and 2 means that the projects engage sources located in foreign countries.

2.5.3.2 Independent Variables

Knowledge novelty The extent of novelty is assessed according to the existing market offerings. For each NPD project we asked respondents to evaluate whether the knowledge generated was new to the industry or only to the firm (thus already present in the world market offerings). From the interviews, it turned out that the novelty of the project was not unexpected by the firm, but there was an ex-ante intent to innovate at a certain level. Indeed, the firms had on the shelf innovative machine concepts in search of industrial applications, but due to the high costs of the machines they decided to further develop the concept only once a client was ready to invest in it. Drawing on innovation literature (Amara et al. 2008; OECD 2005), we measured the degree of project novelty with a dichotomous variable that takes the value 1 when the project introduced knowledge new to the industry and the value 0 when the project introduced knowledge new to the firm. Examples of projects which convey performance features not available in the world market offerings are direct drive heads, without help of gears and belts, which perform rotations both in working and in positioning in a very short time and with unique accuracy. Other examples are systems in press brakes that allow the bending of a sheet at the desired angle in a controlled way with the necessary accuracy without having to go through trial-and-error phases which inevitably lead to waste of material, or a proportional frame deflection compensation system that allows any bend to be made at a constant angle, regardless of the length of the workpiece. These projects are characterized by a high degree of novelty compared to the knowledge embodied in the extant products of the industry. On the other hand, projects which introduce knowledge that is new to the firm but already in the market are, for instance, a plasma cutting system that allows the elimination of the cutting fumes with half the power compared to traditional systems, or "direct drive" rotary tables which grant maximum accuracy and very short rotation times.

Knowledge breadth Prior research measured breadth at the firm level in terms of the expansion of a firm's technology base into a wide range of technological fields (Quintana–Garca and Benavides–Velasco 2008; Zhang et al. 2007; Zhang and Baden–Fuller 2010). At the project level the operationalization of breadth implies the definition of the body of technical knowledge (technological domains), investigated in a specific project, which contributes to solve primary problems through the identification of the operating principles that makes it possible to match functions to components. According to the conceptualization of breadth as the degree of heterogeneity among technological domains, we maintain that the simple count of the technological fields investigated at the project level makes it difficult to assess the complexity that any additional domains bring into a project that already relies on the consolidated body of knowledge of an industry. The breadth measure should take into account the composition of domains which define the technological solution landscape for a specific project. Indeed, a technical problem can be solved by drawing on consolidated disciplines at the base of the sector or can require reliance on additional diverse domains in which the number and the characteristics of the possible alternatives are less defined and more uncertain. Moreover, each additional domain increases the degree of knowledge heterogeneity. Therefore, the breadth in the first case should assume a lower value than in the second case.

In order to operationalize the degree of breadth at the project level, we first drew on the technical characteristics of the machine tool product. A detailed analysis of the sector, based on the review of the specialized literature and discussion with experts, supported the findings of previous studies (Freddi 2009; Mazzoleni 1999; Wengel and Shapira 2004). It turned out that the traditional discipline in the machine tools industry, namely mechanics, is progressively blending with two different bodies of technical knowledge, electronics and software engineering, generating the so-called mechatronic product (Kodama 1992). However, as other research has shown, the core competencies of the machine producer industry still "lie firmly within the mechanical field" (Lissoni 2001, p. 1495).

Each of the three domains solves distinct primary problems. Mechanics offers solutions to problems concerning the acceleration and deformation of objects under known forces or stresses. Operating principles drawn from mechanics allow the transmission of power and movement through racks and ball screws, and the traversing movements by sliding blocks and circulating ball guides. Electronics addresses problems related to the use of the controlled motion of electrons through different media; sub-domains are, for example, control engineering, microelectronics, signal processing. The numerically controlled technology in machine tools is based on the principles of electronics and devices such as drivers, transistors, encoders which allow the movement controls and the automation of processes. Software engineering deals with problems concerning the development of instructions and interfaces for programming and controlling the hardware components. For instance, in machine tools the software automatically creates the programming CAM for the machine for optimizing the working sequence, choosing the right tools and calculating the developments. Thus, this solution from the software domain generates new interdependencies both with mechanical and electronic components.

During the interviews the respondents were asked to indicate which of the three technological domains they investigated to solve technical problems in each project. We calculated the Manhattan distance, comparing the domain composition of each project with that of a "standard" project, which relies exclusively on theconsolidated knowledge domain of the sector (namely mechanics). Accordingly, the degree of breadth has been evaluated on a 3-point scale, where 0 means that there is no distance between a project under scrutiny and a "standard" one, both relying on the single traditional field of the sector (no heterogeneity); 1 means that the domain composition of the project under scrutiny encompasses one technological domain (electronics or software engineering) additional to the single domain of the "standard" project; 2 means that the project encompasses two additional fields (electronics and software engineering). These latter projects are defined exploratory due to their high degree of breadth, since the problem-solving activity implies a search for technical solutions in heterogeneous domains.

2.5.3.3 Control Variables

In accordance with the extant literature on technological innovation we introduced a number of control variables that might influence the propensity of the firm to rely on external sourcing.

We used *firm age* as a proxy for the firm's legacy. As previous studies show, more established firms are more likely to engage in autonomous innovation instead of relying on external actors (Zhao et al. 2005). This might be due to the fact that older firms may have accumulated experience and knowledge over the years, and have built their own in-house capability to become more autonomous in innovation than younger firms. Furthermore, older firms develop established procedures and routines that create resistance to the integration of external sources (Freel 2003; Li and Tang 2010). Given these findings we predict that firm age will have a negative effect on external technology sourcing. Firm age was measured as the number of years since the firm was founded to the year the projects started.

Then we included *firm size* as a proxy for market power. As suggested by empirical research, larger firms have the capacity to attract and to deal with external partners and they are more likely to be engaged in a wider range of activities that may require external sources (Belderbos et al. 2004; Fritsch and Lukas 2001; Tether 2002). Firm size was measured by the logarithm of the average number of the firm's employees over the 2 years before the projects started. Moreover, we measured the ratio of the average number of the firm's R&D employees to the total number of employees over the 2 years before the projects started as a proxy of R&D intensity. Veugelers (1997) found that R&D spending does not have an impact on cooperation in R&D unless firms have their own R&D department and personnel. In previous empirical studies, R&D intensity is used as an indicator of the firm's ability to recognize, value and exploit technological opportunities from outside (Cohen and Levinthal 1990; Fritsch and Lukas 2001). Finally, since our sample is composed of multiple projects from a small number of firms we included dummy variables in the model in order to control for the non-independence of the observation due to firm differences.

Table 2.1 Number of NPD projects by firms

Firms	Internal sourcing	External sourcing	Total number of projects per firm
1	4	5	9
2	8	4	12
3	10	6	16
4	4	7	11
5	4	8	12
6	18	1	19
7	0	7	7
Total number of projects per sourcing decision	**48**	**38**	**86**

2.6 Findings

2.6.1 Open Innovation and Sourcing Project-by-Project

A first preliminary analysis is necessary to understand if in our sample firms adopt a sourcing strategy project-by-project or whether they implement a common strategy across the project portfolio. Indeed, our research hypotheses on the impact of knowledge attributes of an NPD project on sourcing decisions are tested in the first type of firm.

A qualitative analysis carried out on the sample showed that in two firms the decision to draw on internal or external sources in NPD projects is predetermined by a common orientation, namely the protection of knowledge from technological leakage and hold-up risk. From the data we gathered through the field interviews it turned out that one firm prefers to rely on its own R&D resources and on the parent firm, whereas the other develops all the projects through long-term partnerships, which allows a high control over the knowledge generated. The approach adopted by the two companies can be ascribed to the closed innovation paradigm (Chesbrough 2003), which leads to a common sourcing strategy across their NPD projects. On the other hand, in the remaining five firms the interviewees maintained that they assess the two sourcing choices (internal versus external) project-by-project. The qualitative data has been supported by the findings of the independence test described below.

Table 2.1 summarizes the conjoint distribution of the above-mentioned variables: internal versus external sourcing, in the columns, and firms in the rows.

To validate the qualitative analysis on the sample of seven firms we carried out a Chi-square test of independence between the variables firms and internal versus external sourcing decision.

The p-value of the χ^2 test is approximately zero (χ^2 statistic $= 26$, df $= 6$, p-value $= 0.0002$), so we reject the null hypothesis of independence between firms and internal versus external sourcing decision. To verify that the projects of a single

Table 2.2 Statistics and p-value for the leave–1–out χ^2 test for independence

Firm	χ^2 statistics	p–value
1	25.647	0.0001
2	25.152	0.0001
3	25.479	0.0001
4	24.470	0.0002
5	23.672	0.0003
6	10.985	**0.0517**
7	16.932	0.0046

Table 2.3 Statistics and p-value for the leave-2-out χ^2 test for independence

Firm	2	3	4	5	6	7
1	24.897	25.224	23.987	23.132	10.982	16.134
	(0.0001)	(0.0000)	(0.0001)	(0.0001)	(0.0268)	(0.0028)
2		24.433	23.856	23.129	8.399	16.575
		(0.0001)	(0.0001)	(0.0001)	**(0.0780)**	(0.0023)
3			24.178	23.446	8.625	16.841
			(0.0001)	(0.0001)	**(0.0712)**	(0.0021)
4				21.644	10.546	14.347
				(0.0002)	(0.0322)	(0.0063)
5					10.129	13.187
					(0.0383)	(0.0104)
6						4.596
						(0.3313)

firm affect the results of the independence test, we repeated the χ^2 test removing the project of k firms (Bruce and Martin 1989). Therefore, when $k = 1$ we performed seven tests in each of which we dropped the projects of a firm. In the analysis, the testing procedure was conducted for $k = 1$ and $k = 2$. This leave-k-out procedure allows us to exclude the projects of two firms and to consider only the subset of projects of the firms that present potentially analogue behaviour with regard to internal versus external sourcing decisions.

The results of the test leave-1-out are summarized in Table 2.2. They show us that the firm 6 has conditioned the test of independence, indeed the p-value of the independence test when we exclude the data relative to firm 6 is 0.0514. Thus, we accept the null hypothesis of independence.

Table 2.3 summarizes the results of the testing procedure when $k = 2$ and shows us that removing firms 6 and 7 jointly produces a higher p-value (0.3313) than the leave-1-out test. Therefore, we removed the projects of these two firms, which implement a common sourcing strategy across all their NPD projects, and thus we restricted the sample to the projects of those firms which make their sourcing choice project-by-project. This implies that the sample size decreases to 60 NPD projects.

Table 2.4 Cross tabulation of novelty with breadth

Breadth	Novelty		Total
	0	1	
0	11	5	16
1	9	5	14
2	14	16	30
Total	**34**	**26**	**60**

The size of our sample is not small if we compare it to prior studies which adopted the project level of analysis addressing the sourcing decisions in the context of the innovation process. For instance, Cassiman et al. (2010) based their analysis on a sample of 52 R&D projects developed by one company, Kessler et al. (2000) relied upon a survey of 75 NPD projects carried out by ten firms, and Salge et al. (2013) carried out their study on 62 NPD projects developed by one firm.

2.6.2 Hypotheses Testing

The analysis of the composition of the final sample shows that half of the projects drew on external sources, primarily suppliers (24 projects) and, to a lesser extent, universities (5 projects), clients (4 projects), and consultants (4 projects). Concerning the governance modes adopted, 17 out of 60 projects were implemented through R&D development agreements. As far as the geographical distribution of the external sources is concerned, it turned out that 44.4 % of the sources are located beyond national boundaries.

Concerning the knowledge attributes (Table 2.4), according to our definition of *Novelty* the sample is characterized by 43 % of projects new to the industry and by 57 % new only to the firm. The extent of the *Breadth* is equal to 0 in 27 % of the sample, to 1 in 23 % and to 2 in half of the projects.

Table 2.5 displays the descriptive statistics and the correlations of all of the variables included in the model. The χ^2 test carried out between Novelty and Breadth variables shows that the two knowledge attributes are independent (p-value = 0.286). This means that projects with high novelty may require operating principles not necessarily from heterogeneous technological domains, whereas projects with high breadth may convey performance features which may also not be radically new for the industrial demand.

Qualitative cases from our sample support this finding. Exploratory projects from a customer perspective (high novelty) can introduce functions new to the industry by drawing exclusively on mechanical domains. An example can be a project which aims to produce a machine that does not need expensive, bulky and uncomfortable foundation pits, normally necessary in similar machines in the industry to have an acceptable distance between table and spindle nose. The solution to the technical problems, raised by the function required, namely a lowered trim morphology, has

Table 2.5 Descriptive statistics and correlation matrix

	1	2	3	4	5	6	7	8	Variance inflation factor
1. Internal versus external sourcing	1.00								
2. R&D development agreements	**0.63**	1.00							
3. Partner distance	**0.95**	**0.58**	1.00						
4. Novelty	**0.47**	**0.42**	**0.47**	1.00					1.086
5. Breadth	**0.26**	**0.30**	**0.25**	0.19	1.00				1.113
6. Firm age	0.02	– 0.11	0.02	– 0,03	0.00	1.00			1.374
7. Firm size	0.20	– 0.05	0.17	0.18	0.11	**0.36**	1.00		2.176
8. Firm R&D intensity	0.02	0.04	0.00	– 0,12	– 0.25	– 0,08	**– 0.31**	1.00	1.746

Correlation coefficients in bold are significant at 5 % level

been identified investigating exclusively the mechanics domain. On the other hand, exploratory projects from a technical perspective (high breadth) can introduce features new only to the firm as in the case of a project which introduced into a machine direct drive rotary tables, already used by other firms in the industry, in order to achieve higher accuracy and shorter switching times. This project implies a combination of knowledge from the mechanical, the electronics and software engineering fields. Mechanical principles are used for the design of the continuous rotary tables which allow the positioning and the clamping of the pallets on tapers, assuring stability and rigidity during machining operations. The electronics domain is investigated for the implementation of motion and measuring systems (motor, encoder, circuitry and indicator to display actual position and to monitor speed) which guarantee the total absence of backlash and the high resolution direct read-out of the position. Knowledge from the software engineering field means it is possible to program the machine while it is operating.

In order to examine multicollinearity, we calculated the variance inflation factor (VIF). VIFs are all below the rule-of-thumb cut-off of 5, thus issues of multicollinearity do not seem to prompt concern.

To verify the research hypotheses we fitted logit models to the data. This model allows us to use categorical variables. The independent variables are *Novelty*, a dichotomous variable, and *Breadth* whose values range across the following set: 0, 1, 2. The three dependent variables measure the sourcing choices at project level, and specifically:

- internal versus external sourcing, a dichotomous variable, where 1 indicates that the NPD project was carried out drawing on external sources and 0 indicates the absence of external technology sourcing.
- R&D development agreement, a dichotomous variable, where 1 indicates that the project involved R&D agreements for its development and 0 indicates the absence of joint development agreements with external sources.

Table 2.6 Results of the fitting model: internal versus external sourcing in NPD project

Variables	Model 1			Model 2		
	Coefficient	s.e.	*p*–value	Coefficient	s.e.	*p*–value
Constant	– 1.562	0.564	**0.006**	16.360	39.713	0.680
Novelty	1.991	0.610	**0.001**	2.052	0.714	**0.004**
Breadth	0.583	0.330	**0.077**	0.799	0.371	**0.031**
Firm age				0.020	0.203	0.921
Firm size				– 3.562	6.780	0.599
Firm R&D intensity				10.005	41.635	0.810
Dummy for firm 2				– 3.425	4.395	0.436
Dummy for firm 3				– 5.546	7.750	0.474
Dummy for firm 4				– 1.628	4.472	0.716
Dummy for firm 5				– 2.086	4.474	0.641
Chi-square	16.586		**0.000**	20.983		**0.013**
Number of cases	60			60		
Correctly predicted (accuracy rate)	73.3%			78.3%		

- Partner distance, measured on a 3-point scale, where 0 means that the project draws on similar cognitive sources within the firm's boundaries, 1 means that the project involves sources geographically located in the same country, and 2 means that the project engages sources from different countries, and therefore with a higher cognitive distance.

The model for testing H1, H2, H3 and H4 is the following:

$$f(y) = \frac{e^y}{1 + e^y} \tag{2.1}$$

where $f(y)$ may be interpreted as the probability that y is 1 for the external technology sourcing and R&D development agreement. Moreover, y is a linear combination of explanatory variables, that is:

$$y = \beta_0 + \beta_1 \text{Novelty} + \beta_2 \text{Breadth} \tag{2.2}$$

The model has been modified for testing H5 and H6 in order to take into account the dependent variable "partner distance" which is ordinal.

The results of the estimation procedure obtained by gretl software are shown in Tables 2.6, 2.7 and 2.8.

In Tables 2.6 and 2.7 the parameters of the models, that measure the contribution of the independent variables, are all significant and positive. Thus, the result supports H1, H2, H3 and H4, which predict the positive effect of the single knowledge attribute on external sourcing and on R&D development agreements. The two

Table 2.7 Results of the fitting model: R&D development agreement

Variables	Model 3			Model 4		
	Coefficient	s.e.	*p*–value	Coefficient	**s.e.**	*p*–value
Constant	– 3.144	0.833	**0.000**	3.548	54.556	0.948
Novelty	1.899	0.687	**0.005**	2.463	1.153	**0.032**
Breadth	0.859	0.404	**0.033**	1.164	0.444	**0.008**
Firm age				0.101	0.234	0.667
Firm size				– 2.510	9.668	0.796
Firm R&D intensity				12.900	49.789	0.796
Dummy for firm 2				0.922	6.420	0.886
Dummy for firm 3				– 1.197	10.536	0.886
Dummy for firm 4				– 0.476	4.802	0.921
Dummy for firm 5				1.073	4.874	0.826
Chi-square	15.056		**0.000**	20.043		**0.018**
Number of cases	60			60		
Correctly predicted (accuracy rate)	78.3 %			88.3 %		

Table 2.8 Results of the fitting model: partner distance in NPD project

Variables	Model 5			Model 6		
	Coefficient	s.e.	*p*–value	Coefficient	s.e.	*p*–value
Novelty	1.598	0.548	**0.004**	1.604	0.690	0.020
Breadth	0.391	0.330	0.236	0.623	0.369	**0.092**
Firm age				– 0.209	0.196	0.286
Firm size				– 2.676	8.179	0.744
Firm R&D intensity				9.245	37.175	0.804
Dummy for firm 2				– 3.711	5.277	0.482
Dummy for firm 3				– 5.764	8.808	0.513
Dummy for firm 4				– 0.065	3.548	0.985
Dummy for firm 5				– 4.882	4.913	0.320
Chi-square	15.477		**0.000**	20.885		**0.013**
Number of cases	60			60		
Correctly predicted (accuracy rate)	60.0 %			63.3 %		

estimated models seem to be able to correctly predict the sourcing choices: respectively 75 and 88.3 % of the fitted values match the observed values. The *p*-value of the likelihood ratio test is 0.012 and 0.017, thus the models can adequately explain the relationship between variables. Concerning H5 and H6 on the relationship between NPD project's knowledge attributes and the choice to rely on distant partners, the results show that novelty is significant (p-value = 0.02), therefore H5 is supported. As far as knowledge breadth is concerned, the p-value of the estimated coefficient is about 0.09, thus H6 is not supported. Knowledge breadth seems not to impact on decisions to rely on similar versus distant partners. From the results it turned out that firms aiming to add additional technological domains through an NPD project search for this knowledge outside the organizational boundaries and implement R&D development agreements, as confirmed by hypotheses H2 and H4, but the locus of this search can be both national and international, depending on where the specialized knowledge resides. Considering that the machine tool sector has a pivotal role both in the Italian manufacturing system but also in other countries (Germany, USA, Japan), the pool of specialized knowledge for the firms operating in this industry may be both national and international. The estimated model predicts correctly 63.3 % the observed values. The p-value of the likelihood ratio test is 0.013.

Nevertheless, we have to highlight that the number of observations used to estimate the models is small, though sufficient for the models without control variables. This suggests some caution in interpreting the findings.

As regards the control variables, the regression coefficients for the firm age, size and R&D intensity are not significant. The findings also remain robust when introducing the dummies in order to control for firms' differences.

We also investigated the possible effect of the interaction between *Novelty* and *Breadth* on external technology sourcing by estimating a complete model, but we did not find any significant interaction between the two independent variables (the p-value of the parameter of the interaction factor was 0.65).

2.7 Discussion and Conclusion

The chapter has addressed the issue of sourcing determinants proposing a project-level contingency approach. We focused on antecedents at the same level of analysis of the phenomenon under investigation, studying the attributes of the knowledge an NPD project aims to generate. We defined these attributes, accordingly with the literature conceiving the generation of new knowledge by a project as a discovery process within the sectoral knowledge landscape. The first attribute is *knowledge novelty*, defined in relation to the knowledge space of original product concepts and functionalities for market needs, and the second is *knowledge breadth*, defined according to the heterogeneity of technological fields that provides solutions to product problems. In our research we investigated the impact of these two NPD projects' attributes on three key sourcing choices.

Concerning the first decision (external versus internal sourcing), the empirical findings show that when a project engages in exploration at the frontier of knowledge novelty and knowledge breadth external partners are sources from which it might benefit from the point of view of learning advancements, uncertainty reduction and efficiency gains. As far as learning advantages are concerned, external sources, on the one hand, encourage divergent thinking for the generation of new state-of-the-art product concepts and functionalities (high novelty), and on the other hand increase the possibility to accumulate knowledge on heterogeneous technological fields (high breadth). Moreover, external sources might reduce the level of uncertainties related, on the one hand, to market acceptance of new concepts or functionalities when the project aims to meet ahead-of-market needs (high novelty), and on the other hand to the understanding of the possible interdependencies among heterogeneous problem settings (high breadth). Finally, efficiency benefits derive from the preliminary prototyping tests that external sources carry out in implementing new functionalities (high novelty), from the partitioning of project development risks and costs among external specialized partners and finally from compression of time to market due to partner expertise (high breadth).

As far as the second decision (how to source), is concerned results show that the exploration at the frontier of the knowledge space induces firms to involve external partners in NPD projects by means of R&D development agreements. This form of governance of the relationship allows the firm to pursue higher learning benefits in terms of successful identification of far analogies and effective transfer of the similarities detected (high novelty). Moreover, co-development agreements represent a learning vehicle also in terms of better understanding of different technological domain integration and in-house accumulation of specialized knowledge (knowledge breadth).

Concerning the third decision (where to source), from the findings it emerged that knowledge novelty spurs firms to rely on cognitive distant partners. The difficulty to overcome the physical and the cultural barriers seem to be counterbalanced by the benefits a firm can pursue by searching far from its local environment in terms of access to different customer mindsets and discoveries of non-obvious analogies. Especially in the B2B industries, as in the case of machine tools firms, specialized in complex industrial product adapted to the customer's needs with a strong component of complementary services, the search of original products' features and functionalities benefits from the interaction with nonlocal partners that allow the identification of cross-cultural differences that can be included in the firm offering. The exploration of additional technological domains through an NPD project (high knowledge breadth) turned out not to impact on the decision to involve similar rather than distant partners. The decision-making process about this specific sourcing choice can be influenced by further factors such as the relationships that the firm has built in prior NPD projects with partners in specialized domains. If in the past the firm worked with expert sources operating in the electronics or software engineering fields and over time they developed a shared system of meanings and norms, the firm may continue to rely on these sources independently of their geographical location. To this regard, explanatory factors at the project level should be complemented with antecedents at the firm level, such as prior experience with the same partner (Gulati et al. 2009). Moreover, the presence of regional clusters specialized in the same industry in different countries increases the opportunities for the companies to draw on a pool of expert partners within and across national boundaries.

Our research extends and contributes to the literature in three main ways.

First, the paper adds to the inbound open innovation literature by addressing the issue of a firm's sourcing strategy as a portfolio of decisions across NPD projects, complementing studies at the firm level of analysis. Some recent research that focuses at the firm level of analysis has made the implicit assumption that sourcing decisions are made project-by-project (Carson et al. 2006; Huang et al. 2009; Knudsen and Mortensen, 2011), but only a few studies have provided direct evidence of this (Bonesso et al. 2011; Cassiman et al. 2010; Kessler et al. 2000). Building upon this line of enquiry, we were able first to provide empirical evidence of a sample of firms which make their sourcing choices project-by-project, and second to theorize explanatory antecedents at the same level of analysis of sourcing decisions.

Second, we add to the project management literature, drawing attention to the issue of inbound choices related to the NPD project. This research has primarily provided tools and practices that enable firms to increase the effectiveness and efficiency of the internal NPD process neglecting the role played by projects as a means for exploring external sources in open innovation decisions.

Finally, we contributed to the knowledge-based view literature. The focus on the knowledge base determinants at the firm level investigated by previous research has overshadowed the analysis of the knowledge generated at project level (Nickerson and Zenger, 2004). Our research provides a theoretical contribution to overcome this gap by offering a conceptualization and operationalization of the two constructs of novelty and breadth that are coherent with the level of analysis investigated.

Our findings could find generalizability in a number of project-based industries (Hobday 2000), considering the relevance of "the ongoing projectification' of several sectors" (Söderlund et al. 2008, p. 517). Furthermore, we maintain that insights from the sector investigated in our research could be extended to similar industrial systems characterized by technological convergence.

Among the managerial implications that can be drawn from our results, we highlight that sourcing decisions made across a project portfolio calls for the need for a flexible network of collaborations (Faems et al. 2005; van de Vrande 2013) that can be quickly reconfigured in exploratory projects to meet any new market needs and to handle heterogeneous technological domains.

Moreover, our study has provided a point of departure for the debate of sourcing strategy as a portfolio of decisions made across projects. Our research shows that the advantages brought about by external sourcing are contingent to the knowledge a project aims to develop. Therefore, managers could select external sources through a careful analysis of the project portfolio.

Some limitations in this study have to be acknowledged. First, we favoured the richness of the data set on projects built upon a small sample of firms. Replication of the findings in a larger sample would be welcome. Second, we based our analysis on retrospective data. We tried to overcome major bias by the way in which we conducted our interviews (Miller et al. 1997), as reported in the method section; however, the known limits of this data collection method suggest some caution. Moreover, we operationalized partner cognitive distance using the geographical distance as a proxy. This measure does not consider other dimensions of partners' differences in terms of systems of interpretation and meanings, besides the cultural and the physical space factors. Finally, we did not include in our analysis the firms' prior partnering experience, which can complement project-level antecedents in explaining inbound choices.

We suggest that future research takes more explicitly into account the project level of analysis and its attributes in investigating sourcing choices. This research provides the opportunity for some new thoughts about the way knowledge novelty and breadth, as the coordinates of the search space in which a project searches for solutions, can be conceived as well as operationalized. Moreover, given the scarcity of previous theoretical and empirical work on this issue, future research could adopt a multilevel approach and delve into the integration between the knowledge base of the firm and the knowledge attributes of the project as antecedents of boundary spanning choices. Further studies could take into consideration the different firms' sourcing strategies approach adopted (project-by-project or common strategy across the project portfolio) in studying the antecedents that drive sourcing decisions.

References

Al-Laham, A., Amburgey, T. L., (2011). Staying local or reaching globally? Analyzing structural characteristics of project-based networks in German biotech. In Cattani, G., Ferriani, S., Frederiksen, L., Täube, F., (eds) *Project-Based organizing strategic management* (Advances in Strategic Management, Volume 28), (pp. 323–356): Bingley (UK): Emerald Group Publishing.

Amara, N., Landry, R., Becheikh, N., Ouimet, M., (2008). Learning and novelty of innovation in established manufacturing SMEs. *Technovation* 28(7): 450–463.

Asheim, B.T., Coenen L., (2005). Knowledge bases and regional innovation systems: Comparing Nordic clusters. *Research Policy* 34(8): 1173–1190.

Becerra, M., Lunnan, R., Huemer, L., (2008). Trustworthiness, risk, and the transfer of tacit and explicit knowledge between alliance partners. *Journal of Management Studies* 45(4): 691–713.

Bahemia, H., Squire, B., (2010). A contingent perspective of open innovation in new product development projects. *International Journal of Innovation Management* 14(4): 603–627.

Belderbos, R., Carree, M., Diederenc, B., Lokshinb, B., Veugelers, R., (2004). Heterogeneity in R&D cooperation strategies. International Journal of Industrial Organization 22(8–9): 1237–1263.

Bonesso, S., Comacchio, A., Pizzi, C., (2011). Technology sourcing decisions in exploratory projects. *Technovation* 31(10–11): 573–585.

Boschma, R., (2005). Proximity and innovation: a critical assessment. *Regional Studies* 39(1): 61–74.

Bröring, S., Leker, J., (2007). Industry convergence and its implications for the front end of innovation: a problem of absorptive capacity. *Creativity & Innovation Management* 16 (2): 165–175.

Bruce, A.G., Martin R.D., (1989). Leave-k-Out diagnostics for time series. *Journal of the Royal Statistical Society*. Series B (Methodological), 51(3): 363–424.

Brusoni, S., Prencipe, A., (2006). Making design rules: a multidomain *perspective. Organization Science* 17(2): 179–189.

Brusoni, S., Sgalari, G., (2006). New combinations in old industries: the introduction of radical innovations in tire manufacturing. *Journal of Evolutionary Economics* 16(1–2): 25–43.

Brunswicker, S., Hutschek, U. (2010). Crossing horizons: leveraging cross-industry innovation search in the front-end of the innovation process. *International Journal of Innovation Management* 14(4): 683–702.

Burt, R. S., (2005). Brokerage and closure. An introduction to social capital. New York: Oxford University Press.
Carson, S.J., Madhok, A., Wu, T., (2006). Uncertainty, opportunism, and governance: the effects of volatility and ambiguity on formal and relational contracting. *Academy of Management Journal* 49(5): 1058–1077.
Cassiman, B., Veugelers, R., (2006). In search of complementarity in innovation strategy: internal R&D and external knowledge acquisition. *Management Science* 52(1): 68–82.
Cassiman, B., Di Guardo, M.C., Valentini, G., (2010). Organizing links with science: cooperate or contract? A project-level analysis. *Research Policy* 39(7): 882–892.
Chen, L.C., (2009). Learning through informal local and global linkages: the case of Taiwan's machine tool industry. *Research Policy* 38(3): 527–535.
Chesbrough, H.W., (2003). *Open innovation: the new imperative for creating and profiting from technology*. Boston: Harvard Business School Press.
Chesbrough, H.W., (2006). *Open business models: How to thrive in the new innovation landscape*. Boston: Harvard Business School Press.
Chiaroni D., Chiesa V., Frattini F., (2010). Unravelling the process from closed to open innovation: evidence from mature, asset intensive industries. *R& D Management* 40(3): 222–245.
Choi, D., Valikangas, L., (2001). Patterns of strategy innovation. *European Management Journal* 19(4): 424–429.
Cohen, W.M., Levinthal, D.A., (1990). Absorptive capacity: a new perspective on learning and innovation. Administrative Science Quarterly 35(1): 128–152.
Colarelli O'Connor, G., (1998). Market learning and radical innovation: a cross case comparison of eight radical innovation projects. *The Journal of Product Innovation Management* 15(2):151-66.
Cummings, J.L., Holmberg, S.R., (2012). Best-fit alliance partners: the use of critical success factors in a comprehensive partner selection process. *Long Range Planning* 45(2–3): 136–159.
Dahl, D.W., Moreau, P., (2002). The influence and value of analogical thinking during new product ideation. *Journal of Marketing Research* 39(1): 47–60.
Danneels, E., Kleinschmidt, E.J., (2001). Product degree of novelty of innovation from the firm's perspective: its dimensions and their relation with project selection and performance. *The Journal of Product Innovation Management* 18 (6): 357–373.
Dittrich, K., Duysters, G., (2007). Networking as a means to strategy change: the case of open innovation in mobile telephony. *The Journal of Product Innovation Management* 24(6): 510–521.
de Jong J.P.J., von Hippel, E., (2009). Transfers of user process innovations to process equipment producers: a study of Dutch high-tech firms. *Research Policy* 38(7): 1181–1191.
Enkell, E., Gassmann, O., (2010). Creative imitation: exploring the case of cross-industry innovation. *R&D Management* 40(3): 256–270.
Faems, D., Looy, B.V., Debackere, K., (2005). Interorganizational collaboration and innovation: toward a portfolio approach. *The Journal of Product Innovation Management* 22(3): 238–250.
Fleming, L., Sorenson, O., (2003). Navigating the technology landscape of innovation. *MIT Sloan Management Review* 44 (2): 15–23.
Freel, M.S., (2003). Sectoral patterns of small firm innovation, networking and proximity. *Research Policy* 32(5): 751–770.
Freel, M., de Jong, J.P.J., (2009). Market novelty, competence-seeking and innovation networking. *Technovation* 29(12): 873–884.
Freddi, D., (2009). The integration of old and new technological paradigms in low- and medium-tech sectors: the case of mechatronics. *Research Polic* 38(3): 548–558.
Fritsch, M., Lukas, R., (2001). Who cooperates on R&D? *Research Policy* 30(2): 297–312.
Gambardella, A., Torrisi, S., (1998). Does technological convergence imply convergence in markets? Evidence from the electronics industry. *Research Policy* 27(5): 445–463.
Garriga, H., von Krogh, G., Spaeth, S., (2013). How constraints and knowledge impact open innovation. *Strategic Management Journal* 34(9): 1134–1144.

Gassmann, O., Zeschky, M., (2008). Opening up the solution space: the role of analogical thinking for breakthrough product innovation. *Creativity and Innovation Management* 17(2): 97–106.
Gassmann, O., Enkel, E., Chesbrough, H. (2010). The future of open innovation *R &D Management* 40(3): 213–221.
Gentner, D. (1983). Structure-mapping: a theoretical framework for analogy. *Cognitive Science* 7(2): 155–170.
George, G., Kotha, R., Zheng, Y., (2008). Entry into insular domains: A longitudinal study of knowledge structuration and innovation in biotechnology firms. *Journal of Management Studies* 45(8): 1449–1474.
Gulati, R., Lavie, D., Singh, H., (2009). The nature of partnering experience and the gains from alliances. *Strategic Management Journal* 30(11): 1213–1233.
Hagardon, A., Sutton, R. (1997). Technology brokering and innovation in a product development firm. *Administrative Science Quarterly* 42(4): 716–749.
Henderson, R. M., Clark, K.B., (1990). Architectural innovation: the reconfiguration of existing product technologies and the failure of established firms. *Administrative Science Quarterly* 35(1): 9–30.
Hobday, M., (2000). The project-based organization: an ideal form for managing complex products and systems? *Research Policy* 29(7–8): 871–893.
Hsieh K. N., Tidd, J., (2012). Open versus closed new service development: the influences of project novelty. *Technovation* 32(11): 600–608.
Huang, Y., Chung, H., Lin, C.H., (2009). R&D sourcing strategies: determinants and consequences. *Technovation* 29(3): 155–169.
Kalogerakis, K., Lüthje, C., Herstatt, C., (2010). Developing innovations based on analogies: experience from design and engineering consultants. *The Journal of Product Innovation Management* 27(3): 418–436.
Katila, R., Ahuja, G., (2002). Something old, something new: a longitudinal study of search behaviour and new product introduction. *Academy of Management Journal* 45(6): 1183–1194.
Kessler, E.H., Bierly, P.E., Gopalakrishnan, S., (2000). Internal vs external learning in new product development: effects on speed, costs and competitive advantages. *R&D Management* 30(3): 213–223.
Knudsen, M.P., Mortensen, T.B., (2011). Some immediate-but negative-effects of openness on product development performance. *Technovation* 30(1): 54–64.
Kodama, F., (1992). Technology fusion and the new R&D.*Harvard Business Review July-August* 70-78.
Laursen, K., Salter, A., (2006). Open for innovation: The role of openness in explaining innovation performance among U.K. manufacturing firms. *Strategic Management Journal* 27(2): 131–150.
Laursen, K., Leone, M.I., Torrisi, S. (2010). Technological exploration through licensing: new insights from the licensee's point of view. *Industrial and Corporate Change* 19(3): 871–897.
Lenfle, S., (2008). Exploration and project management. *International Journal of Project Management* 26(5): 469–478.
Li, H.-L., Tang, M.-J., (2010). Vertical integration and innovative performance: the effects of external knowledge sourcing modes. *Technovation* 30(7–8): 401–410.
Lissoni, F., (2001). Knowledge codification and the geography of innovation: the case of Brescia mechanical cluster. *Research Policy* 30(9): 1479–1500.
Macher, J.T., (2006). Technological development and the boundaries of the firm: a knowledge-based examination in semiconductor manufacturing. *Management Science* 52(6): 826–843.
MacPherson, A.D., Kalafsky, R.V., (2003). The technological revitalization of a mature US industry. The case of machine tools. *The Industrial Geographer* 1(1): 16–34.
Malerba, F., (2002). Sectoral systems of innovation and production. *Research Policy* 31(2): 247–264.
Malerba, F., (2004). *Sectoral systems of innovation: concepts, issues and analyses of six major sectors in Europe*. Cambridge: Cambridge University Press.

Malerba, F., (2005). Sectoral systems of innovation: a framework for linking innovation to the knowledge base, structure and dynamics of sectors. *Economics of Innovation and New Technology* 14(1–2): 63–82.
Marsili, O., Salter, A., (2005). Inequality of innovation: skewed distributions and the returns to innovation in Dutch manufacturing. *Economics of Innovation and New Technology* 14(1–2): 83–102.
Mazzoleni, R., (1999). Innovation in the MT industry: a historical perspective on the dynamics of comparative advantage. In Mowery, D., Nelson, R. (eds), *Sources of Industrial Leadership: Studies of Seven Industries.* (pp. 169–216). New York: Cambridge University Press.
McEvily, B., Zaheer, A., (1999). Bridging ties: a source of firm heterogeneity in competitive capabilities. *Strategic Management Journal* 20(12): 1133–1156.
Miller, C., Cardinal, L., Glick, W., (1997). Retrospective reports in organizational research: a reexamination of recent evidence. *Academy of Management Journal* 40(1): 189–204.
Mol, M.J., (2005). Does being R&D intensive still discourage outsourcing? Evidence from Dutch manufacturing. *Research Policy* 34(4): 571–582.
Mortara L., Minshall T., (2011). How do large multinational companies implement open innovation? *Technovation* 31(10–11): 586–597.
Mowery, D.C., Oxley, E.J., Silverman, B.S., (1998). Technological overlap and inter-firm cooperation: implication for the resource-based view of the firm. *Research Policy* 27(5): 507–523.
Narula, R., Hagedoorn, J., (1999). Innovating through strategic alliances: moving towards international partnerships and contractual agreements. *Technovation* 19(5): 283–294.
Nickerson, J.A., Zenger, T. R., (2004). A knowledge-based theory of the firm-The problem-solving perspective. *Organization Science* 15(6): 617–632.
Nooteboom, B., Haverbeke, W.V., Duysters, G., Gilsing, V., van den Oord, A., (2007). Optimal cognitive distance and absorptive capacity. *Research Policy* 36(7): 1016–1034.
OECD, (2005). *Oslo Manual: proposed guidelines for collecting and interpreting technological innovation data.* Third revisited edition. Paris, OECD.
Parida, V., Westerberg, M., Frishammar, J., (2012). Inbound open innovation activities in High-Tech SMEs: the impact on innovation performance. *Journal of Small Business Management* 50(2): 283–309.
Pavitt, K., (1984). Sectoral patterns of technical change: towards a taxonomy and a theory. *Research Policy* 13(6): 343–373.
Podsakoff, P.M., Organ, D., (1986). Self-reports in organizational research: problems and prospects. *Journal of Management* 12(4): 531–544.
Quintana-García, C., Benavides-Velasco, C.A., (2008). Innovative competence, exploration and exploitation: The influence of technological diversification. *Research Policy* 37(3): 492–507.
Robertson, T.S., Gatignon, H., (1998). Technology development mode: a transaction cost conceptualization. *Strategic Management Journal* 19(6): 515–531.
Robertson, P.L., Pol, E., Carroll, P., (2003). Receptive capacity of established industries as a limiting factor in the economy's rate of innovation. *Industry and Innovation* 10(4): 457–474.
Rothaermel, F.T., Boeker, W., (2008). Old technology meets new technology: complementarities, similarities, and alliances formation. *Strategic Management Journal* 29(1): 47–77.
Salge, T.O., Farchi, T., Barrett, M.I., Dopson, S., (2013). When does search openness really matter? A contingency study of health-care innovation projects. *The Journal of Product Innovation Management* 30(4): 659–676.
Sandven, T., Pedersen, T.E., Smith, K., (2001). Analysis of CIS data on the impact of innovation on growth in the sector of machinery and equipment and of electrical machinery. Step Group, Oslo.
Seidel, V.P., (2007). Concept shifting and the radical product development process. *The Journal of Product Innovation Management* 24(6): 522–533.
Simon, H.A., (1962). The architecture of complexity. *Proceedings of the American Philosophical Society* 106(6): 467–482.
Smith, W.K., Tushman, M.L., (2005). Managing strategic contradictions: a top management model for managing innovation streams. *Organization Science* 16(5): 522–536.

Söderlund, J., Vaagaasar, A.L., Andersen, E.S., (2008). Relating, reflecting and routinizing: Developing project competence in cooperation with others. *International Journal of Project Management* 26 (5): 517–526.
Sonali, K.S., Corley, K.G., (2006). Building better theory by bridging the quantitativequalitative divide. *Journal of Management Studies* 43(8): 1821–1835.
Teixeira, A.A.C., Santos, P., Brochado, A.O., (2008). International R&D Cooperation between Low-tech SMEs: The role of cultural and geographical proximity. *European Planning Studies* 16(6): 785–810.
Terwiesch, C., Xu, Y., (2008). Innovation contests, open innovation, and multiagent problem solving. *Management Science* 54(9): 1529–1543.
Tether, B., (2002). Who co-operates for innovation, and why: an empirical analysis. *Research Policy* 31(6): 947–967.
Tranekjer, T.L., Søndergaard, H.A. (2013). Sources of innovation, their combinations and strengths - benefits at the NPD project level. *International Journal of Technology Management* 61(3–4): 205–236.
Trombini G., Comacchio A., (2012). Cooperative markets for Ideas: when does technology licensing combine with R&D partnerships?, *DRUID Conference*, Copenhagen Business school, Copenhagen; Denmark, 19–21 June 2012.
Tsai, K-H., (2009). Collaborative networks and product innovation performance: toward a contingency perspective. *Research Policy* 38(5): 765–778.
Ucimu, (2006). *Annual report*. Milano: Ucimu.
Veugelers, R., (1997). Internal R&D expenditures and external technology sourcing. *Research Policy* 26(3): 303–315.
Veugelers, R., Cassiman, B., (1999). Make and buy in innovation strategies: evidence from Belgian manufacturing firms. *Research Policy* 28(1): 63–80.
van de Vrande, V., (2013). Balancing your technology-sourcing portfolio: how sourcing mode diversity enhances innovative performance. *Strategic Management Journal* 34(5): 610–621.
van de Vrande, V., Vanhaverbeke, W., Duysters, G., (2009). External technology sourcing: The effect of uncertainty on governance mode choice. *Journal of Business Venturing* 24(1): 62–80.
Wang, Q., von Tunzelmann, G.N., (2000). Complexity and the function of the firm: breadth and depth. *Research Policy* 29(7–8): 805–818.
Wengel, J., Shapira, P., (2004). Machine tools: the remaking of a traditional sectoral innovation system. In: Malerba F. (eds), Sectoral *systems of innovation. Concepts, issues and analysis of six major sectors in Europe*, (pp. 243–286). Cambridge: Cambridge University Press.
Zajac E.J., Olsen, C.P., (1993). From transaction cost to transactional value analysis: implications for the study of interorganizational strategies. *Journal of Management Studies* 30(1): 131–145.
Zhang, J., Baden-Fuller, C., (2010). The influence of technological knowledge base and organizational structure on technology collaboration. *Journal of Management Studies* 47(4): 679–704.
Zhang, J., Baden-Fuller, C., Mangematin, V., (2007). Technological knowledge base, R&D organization structure and alliance formation: evidence from the biopharmaceutical industry. *Research Policy* 36(4): 515–528.
Zhao, H., Tong, X., Wong, P.K., Zhu, J., (2005). Types of technology sourcing and innovative capability: An exploratory study of Singapore manufacturing firms. *Journal of High Technology Management Research* 16(2): 209–224.

Chapter 3
A Project-Based Perspective on Complex Product Development

Markus Becker, Luisa Errichiello and Francesco Zirpoli

Abstract In this chapter we review the literature on complex product development focusing on a project-based perspective. We start from showing the specific nature of complex product development processes, and acknowledge the need for relying on external sources of innovation and evaluating its organizational implications. We then focus on the challenges of leveraging such dispersed knowledge, pointing to the specific problems brought by the crucial role of "learning by doing" in complex product innovation processes. The chapter highlights the necessity of shifting the focus of attention from firms' knowledge boundaries to the project knowledge boundaries, so as to gain a more fine-grained analysis of some important phenomena that happen "around" the formal boundary of the firm and cope with knowledge development problems. In the conclusion we hint at the necessity to investigate in more depth *how* using development projects as unit of analysis can contribute to offering new ways of performing organizational ambidexterity.

M. Becker (✉)
Strategic Organization Design Unit, University of Southern Denmark, Campusvej 55, 5230 Odense M, Denmark
e-mail: mab@sod.dias.sdu.dk

L. Errichiello
Institute for Service Industry Research (IRAT), Italian National Research Council (CNR), Via Guglielmo Sanfelice, 8, 80134 Napoli, Italy
e-mail: l.errichiello@irat.cnr.it

F. Zirpoli
Department of Management, Ca' Foscari University of Venezia, San Giobbe Cannaregio, 873, 30121 Venezia, Italy
e-mail: fzirpoli@unive.it

S. Bonesso et al. (eds.), *Project-Based Knowledge in Organizing Open Innovation*,
DOI: 10.1007/978-1-4471-6509-5_3,

3.1 The Role of "Projects" in Complex Products Development

Innovation is typically organized in the form of projects, i.e., temporary organizational structures specifically aimed at developing new products and services. The innovation management literature has recognized that "projects" drive business innovation and change (e.g. Shenhar and Dvir 2007) and that focusing on the "project" level is crucial to managing new product development (e.g. Wheelwright and Clark 1992a).

Similar to what happened to the broader organizational research on projects and project management (Engwall 2003), innovation management and product development scholars have viewed "projects" in two different ways. Among a first group of studies, the project simply provides the empirical setting to investigate a specific phenomenon of interest to innovation scholars. This is, for example, the case of boundary management activities (e.g. Ancona and Caldwell 1990), teamwork (e.g. Pinto and Pinto 1990), knowledge creation and information transfer (e.g. Clark and Fujimoto 1991), or innovation performance outcomes (e.g. Kessler et al. 2000).

In another group of studies, the "project" dimension is put in the foreground and the analytical focus is, for example, on the innovation processes in project-based organizations (Brusoni et al. 1998; Gann and Salter 2000) and the project as the core unit of analysis for understanding: (1) the development of complex product systems, or "CoPS" (Davies and Brady 2000; Hobday 2000), (2) the influence of project management structures on product development success (e.g. Larson and Gobeli 1989; Tatikonda and Rosenthal 2000), and (3) the role of multi-project thinking for effective product development (e.g. Cusumano and Nobeoka 1998; De Maio et al. 1994; Wheelwright and Clark 1992b).

In dealing with the management of innovation within firms producing complex products and systems (such as design, engineering, and construction), Brusoni et al. (1998) pointed to the fact that these firms are always organized around projects: project processes exist outside the traditional firm boundaries, since these firm operate in a multi-actor environment where the project acts as a coordination mechanism across a number of participating firms. Accordingly, in addressing the challenges of organizing innovation processes in CoPS firms primary attention should be given to the project level unit of analysis (Hobday 2000).

Innovation studies focused on the relation between projects and product development have provided valuable contribution to exploring the influence of project-related dimensions (such as project management structures, integration of functional aspects of projects, etc.) on the success of product development activities (e.g. Karlsson and Ahlström 1996; Larson and Gobeli 1989). However, this research seems to suffer from the so-called "lonely project" syndrome (Engwall 2003, p. 790) not only because the main focus remains on individual development projects but also because "the space at which 'innovation' and 'projects' comes together is still dominated by ideas on how to correctly manage projects, rather than how to effectively manage innovation" (Keegan and Turner 2002, p. 367).

Such a narrow perspective has been successfully overcome by a few innovation scholars who contributed to shifting the focus from isolated projects pointing to the key role of multi-project management in new product development activities and the relationship between projects and the organization as a whole (e.g. Cusumano and Nobeoka 1998; De Maio et al. 1994; Leonard-Barton 1992; Wheelwright and Clark 1992b). In these studies the issues under investigation concern project portfolio selection, strategy, and coordination since it is assumed that multi-project management tools affect firm's capabilities and product development performance. Leonard–Barton (1992), in particular, was among the first innovation scholars who underlined the need for overcoming a perspective on new product projects as "self-contained units of analysis" and to study the interface between the project and the organization in general and the systematic interdependences between development practices and firm's core capabilities in particular (Leonard-Barton 1992, p. 112). More specifically, embracing the "lens" of the project the empirical research carried on by Leonard–Barton (1992) shows how development projects can over time foster organizational change through the introduction of new capabilities and the disruption of core rigidities. Wheelwright and Clark (1992a) go further in this direction focusing on the timing of projects and the choice of a variety of project types. In particular, they show how sequencing projects to alternate between different project types can extend employees' skills, help to identify weaknesses in capabilities, improve development processes, and sustain the integration of new tools and techniques into the organization of new product development. De Maio et al. (1994) offer a framework for successfully applying project management techniques to new product development addressing the need to manage project interdependencies while assuring their mutual compatibility at portfolio level. Finally, in their famous book, Cusumano and Nobeoka (1998) document how leading companies in the automotive industry (such as Toyota, General Motors and Fiat) have shifted their attention beyond the efficient management of individual projects in order to optimize product development for the good of the firm as a whole. They point, in particular, to special organizational mechanisms and processes to exploit project interdependencies and enable the sharing of common core components, tasks, and human resources among different projects. "Concurrent design transfer," for example, is a notable practice that can be used by a company to reuse technology from a base project for another and conduct joint design work (Nobeoka 1995; Nobeoka and Cusumano 1997). Indeed, knowledge transfer across new product development projects is the essence of multi-project management and relies on specific inter-project learning mechanisms (Nobeoka 1995; Prencipe and Tell 2001).

3.2 Managing Projects for Distributed Innovation Processes

From the brief literature review provided above, it clearly emerges that innovation management scholars have gradually acknowledged—more or less explicitly—the central role of "projects" in innovation processes and in particular in new product development activities. However, the mentioned literature suffers from some important limitations. First, the literature on project-based organizing in CoPS firms tends to mark the differences between "development projects" and "capital projects" (Winch 1997), pointing to the fact that while the former are often conducted in-house and are concerned with the integration of new technologies into a product (e.g. Iansiti 1995; Iansiti and Clark 1994), "capital projects" (i.e., the case of CoPS) involve a number of external actors and both development and implementation (i.e., manufacturing) are organized on a project basis (Davies and Brady 2000). Making such a marked distinction is not straightforward in light of the increasing diffusion of distributed models of innovation in a number of industries and firms (Chesbrough 2003; Powell et al. 1996; see Prencipe 2003 for a literature review on innovation networks as organizational forms). In particular in the context of new product development, it has become common and, in fact, usual, to organize product development in projects where external sources of innovation are involved (Clark 1989; von Hippel 1988; Nishiguchi 1994; Rothwell 1992). As a consequence of the adoption of new organizational forms of innovation, development projects (and not only capital projects of CoPS) also cross firm boundaries so that boundaries of development projects and firm boundaries do not necessarily coincide any longer. Secondly, in a similar way, scholars addressing the role of projects and project management in new product development have emphasized the relevance of the project as a powerful unit of analysis to address some crucial issues, namely knowledge creation, learning, and competence development. However, the focus remains exclusively on the relationship between projects and the whole organization and it is assumed that projects boundaries fall within those of the innovating firm: the consequences of the open nature of development processes that makes the project the "site" where knowledge creation and learning opportunities are allocated among internal and external participants in the development process are substantially neglected. Finally, the innovation management literature focused on new product development has recently called attention to the importance of considering the "project" level for shedding light on some crucial issues inherent to the adoption of open innovation forms (Bahemia and Squire 2010; Bonesso et al. 2011; Cassiman et al. 2010; Hoang and Rothaermel 2010; Hsieh and Tidd 2012; Salge et al. 2013). As Bahemia and Squire (2010, p. 604) point out, many studies "make the implicit assumption that the choice of an open innovation strategy is determined at the level of the organization and do not provide advice on the degree to which managers should engage external parties during the course of any single (NPD) project. Arguably, such a micro level of analysis may be beneficial where the objectives and conditions of individual NPD projects vary substantially within the same organization, each demanding a different degree of openness." Moreover, focusing on the project level of analysis rather than the firm level offers the chance of a more subtle understanding of learning processes within contexts of R&D collaboration with external actors as well as the relationship between the project and the development of internal capabilities (Hoang and Rothaermel 2010, p. 735). Finally, the "project" constitutes a fine-grained level of investigation that enhances the understanding of the antecedents of sourcing decisions at the level of single projects and can be very useful to empirically investigate how project characteristics influence R&D sourcing decisions (Bonesso et al. 2011). In the next section we delve into the specific case of the development processes of complex products. The following paragraphs show how the specific nature of such products makes the leverage of external sources of innovation particularly complex and posits a number of challenges in organizing their development across the firm boundaries with specific regard to the leverage of dispersed knowledge and the crucial role of learning by doing.

3.3 Key Challenges in Organizing Open Innovation: The Case of Complex Product Development

Products such as cars, PCs, airplanes, or software are complex[1] systems in that they comprise a large number of components and subsystems with many interactions between them. The scheme by which a product's functions are allocated to its components is called product "architecture" (Ulrich 1995). For these products, the allocation of design and engineering activities to external sources of innovation is for OEMs[2] often a necessity rather than a potential choice, since their development often requires the contribution and integration of unique knowledge inputs. For instance, in developing a new model of a car, chassis, engine, interior, and many components are involved, as is market intelligence about customers, knowledge about new materials, and knowledge about new production technology, to give only a few examples. A high degree of specialization is required in all cases and specialized knowledge and competences are often located outside the firm's boundaries. Indeed, most of the operational benefits (cost, quality, and lead times) linked to the involvement of suppliers in the OEMs' NPD processes are due to their specialization on component and system technologies. Making use of specialized suppliers is particularly important in complex multi-technology products because due to the multi-technology and multi-component nature of products, firms cannot maintain all the relevant knowledge bases in-house (Brusoni et al. 2001; Prencipe 2003). Furthermore, suppliers' involvement is often unavoidable because part of their specialized knowledge is tacit and thus difficult and time-consuming to replicate (Dierickx and Cool 1989; Nonaka and Takeuchi 1995; Winter and Szulanski 2001).

From a management perspective, two central challenges arise in the organization of complex NPD processes across firm's boundaries: on the one hand firms need to establish how and to what extent to allocate design and engineering activities to external sources (Brusoni et al. 2001; Takeishi 2001); on the other hand they

[1] For the arguments in this section, see also Zirpoli and Becker (2011a, b) in which we draw on our empirical work carried out in the automotive industry.

[2] Original equipment manufacturers. In the automotive industry, OEM companies include, for example, General Motors, Ford, and Toyota. In some industries, such as electronics, OEMs build products or components used in products sold by another company (often called a value-added reseller, or VAR), in others they are identified as ODM (Original Design Manufacturer). Here, we refer to the OEM as the leader in its value chain, i.e., as the final system integrator.

have to find appropriate organizational mechanisms to integrate and coordinate the specialist knowledge and competences owned by all participants (especially external) in the NPD process (e.g. Brusoni and Prencipe 2006; Zirpoli and Becker 2011a). Actually, the two choices are highly interdependent since outsourcing the development of critical product components to external suppliers can lead to the "hollowing out" of the firm's knowledge base that is crucial to integrate the complex knowledge dispersed in the value chain, and thus, also outside the firm. Integrating specialized knowledge inputs requires that the OEMs hold enough specific knowledge about components but also the "architectural knowledge" that enables a firm to identify the components of a product, the way they are integrated into a system and 'how to coordinate' them (Takeishi 2002, p. 321). Lacking "systemic" knowledge produces, in fact, poor system performance of the product and requiring redesign or reintegration of components and subsystems, leads to additional costs and longer delivery times. In addition, as a consequence of the loss of such systematic knowledge, firms also have problems in specifying and evaluating components, in identifying qualified bidders, evaluating bids, verifying that items meet specifications, improving bids, helping suppliers technically or operationally, improving items after receipt, or finally, being able to make in-house (Fine and Whitney 1996; Lincoln et al. 1998). In turn, this implies serious problems in coordinating and governing suppliers (Helper et al. 2000).

In the case of complex products, their architecture influences to what extent the above challenges can be crucial and particularly difficult to address. Complex products, in fact, typically comprise a mix of components, some of which are tightly coupled to others and some of which are relatively independent (Campagnolo and Camuffo 2010; Salvador 2007). When product architecture is "modular" (Baldwin and Clark 2000; Sanchez and Mahoney 1996), the product can be decomposed in components (i.e., "modules") characterized by interdependence within modules and independence between them (examples are personal computers or bicycles). In this case, allocating design and engineering tasks among NPD participants does not require complex evaluations and the integration of dispersed and highly specialized knowledge is simplified by the existence of standard interfaces among modules (Baldwin and Clark 2000; Sturgeon 2002). On the contrary, for products like cars, owing to their "integral" architecture (MacDuffie 2013; Takeishi and Fujimoto 2003), the main difficulty arises from the functioning of the product as a system. Indeed there are serious limits to how fully one can specify how much an individual component or system contributes to the product performance as a whole because complex technical interdependencies exist between individual components (Zirpoli and Becker 2011b). In this case the isomorphism between product architecture (task decomposition scheme) and the allocation of innovation tasks (and of the related component-specific knowledge) to suppliers does not apply. Moreover, even when product architectures are predictable, the product technology could change and with it also the interdependencies existing between components. This means that how tasks are partitioned should be modified and adapted to technology evolution over time (Brusoni et al. 2001; Chesbrough and Kusunoki 2001; Takeishi 2002).

Also, for integral products firms need to involve external sources of innovation since they often do not have all the resources required to develop in-house all the technologies required to the design and engineering of the product (Christensen 2006). At the same time, since complex products often contain multi-technology components characterized by uneven rates of change of component technologies (Brusoni et al. 2001), drawing on the competences of specialized suppliers is necessary for being able to use the latest technologies (Weigelt 2009). In both cases, OEMs need to carefully assess the impact of their task partitioning choices on the evolution of their component-specific knowledge and architectural knowledge.

In the next paragraph we will show how in addressing the challenges of organizing open innovation for developing complex products the innovation management literature has put a strong emphasis on firm boundaries when the link between task partitioning (who does what) and knowledge partitioning (who knows what) cannot be simply managed by relying on modular product architectures, i.e., as a mechanistic consequence of a given product decomposition scheme. Such a focus has contributed to making vertical scope the most appropriate unit of analysis for establishing task allocation decisions and their linkages with the division of knowledge among NPD participants.

3.4 Addressing the Challenges of Complex Product Development: System Integration, Knowledge Partitioning and the Key Role of Learning by Doing

In addressing the typical situation of complex product innovation, namely that product architectures are integral and/or technologies are complex, heterogeneous, and with an uneven rate of change, the most widely accepted approach is based on the concept of "system integration" (Brusoni and Prencipe 2001; Frigant and Talbot 2005; Mikkola 2006; Miozzo and Grimshaw 2005; Staudenmeyer et al. 2005). Systems integrators are firms that 'bring together high-technology components, subsystems, software, skills, knowledge, engineers, managers, and technicians to produce a product in competition with other suppliers' (Hobday et al. 2005, p. 1110). This requires 'design[ing] and integrat[ing] systems, while managing networks of component and subsystem suppliers' (Hobday et al. 2005, p. 1109).

In conditions of unpredictable interdependencies and uneven rates of change, system integration is supposed to show superior coordination properties when compared to market mechanisms and vertical integration (Brusoni et al. 2001). This argument has therefore pitched systems integration as a third alternative to firms and markets. In order to achieve strong systems integration capabilities (i.e., becoming competent in implementing systems integration), Brusoni and Prencipe (2001) sustain, firms need to "know more than they make." In other words, firms should maintain a knowledge base that is broader than the immediate knowledge required for producing the goods and services (for instance, extending to knowledge of underlying technologies). The authors show how firms that produce aircraft control systems actively engage in R&D

and patent in the areas of underlying technologies used to realize their products (Brusoni and Prencipe 2001; for similar evidence see Nesta and Dibiaggio 2003). Having a broader knowledge base allows system integrator firms to "cope with imbalances caused by uneven rates of development in the technologies they rely on, and with unpredictable product-level interdependencies" (Brusoni and Prencipe 2001, p. 597). The integration of components and subsystems requires that firms stay "on the safe side" by retaining some of the underlying (component-specific) knowledge (Brusoni et al. 2001; Lincoln et al. 1998; Takeishi 2001, 2002).[3] Notwithstanding the importance of some component-specific knowledge, the building block of system integration competences remains, however, "architectural knowledge." As observed by Hobday et al. (2005, p. 1128), in the case of system integration capabilities, "the lead firm moves away from an in-depth control over component design and manufacturing to the systems integration knowledge and skills needed to integrate the modules produced by others in the supply chain."

By emphasizing the alignment of firms' tasks and competence boundaries as a crucial design variable this literature has strongly shifted the focus on firm boundaries and puts in the foreground questions related to the innovating firms' boundaries decisions. The excessive emphasis on architectural knowledge, i.e., the "systems integration knowledge" (Hobday et al. 2005) has contributed to moving the attention away from the crucial role that component-specific knowledge plays to secure system integration capabilities. Indeed, when the goal is to acquire and retain some component-specific knowledge a basic mechanism that cannot be easily replaced is learning by doing (Argote 1999; Argote and Epple 1990). Actually, when a firm decides to outsource all design and engineering tasks related to specific components to external suppliers it is depriving itself of a precious mechanism for acquiring new competences, improving existing ones, or even simply maintaining existing competence levels. The innovation literature has emphasized how the lack of learning by doing can weaken a firm's absorptive capacity (Cohen and Levinthal 1990), the consequent ability to understand technology development and react appropriately by integrating external knowledge into its own product development process. In the case of complex product development, learning by doing as a key mechanism to develop specific knowledge about components is particularly important when such specialized knowledge is tacit or difficult to codify (Kogut and Zander 1992; Nonaka and Takeuchi 1995; Polanyi 1966) and is acquired mainly through its application (Orlikowski 2002). Developing specific knowledge about product components directly contributes to building architectural knowledge and increases the capability to follow the evolution in the product architecture caused by technological evolutions in its components (Takeishi 2001, 2002). Beyond depriving the firm of valuable

[3] Empirical evidences show that firms that pushed the balance of architectural versus component-specific knowledge to the extreme limit by trying to focus on architectural knowledge and outsourcing as much component-specific knowledge as possible, are reported to have problems (Fine and Whitney 1996; Lincoln et al. 1998). In turn, this implies serious problems in coordinating and governing suppliers, leading to an increasing risk of dependency on suppliers (Fine 1998).

competences (such as design competences), the lack of learning by doing can also contribute to reducing its relationship capabilities. Indeed, an excessive knowledge gap about components between OEMs and their suppliers can increase the risks of governance problems (Lincoln et al. 1998; Fine 1998; Takeishi 2001). From the above discussion it clearly emerges that Brusoni et al. (2001) argument, i.e., the idea that firm needs to "know more than they make," should be integrated with the concept that in pursuing this goal OEM firms necessarily have to "make." In the context of NPD, "make" necessarily implies that firms need to be actively involved in at least some development design and engineering tasks of specific components.

The fact that previous literature has chosen the whole firm as the main focus of analysis in the context of the development of complex products has hither to impeded to gain a more fine-grained analysis of some important issues that the firm should take into account when it has to leverage external sources of innovation in developing complex products. In the following paragraphs we show how shifting the focus of analysis from the firm level to the "project" level can provide a fresh and fine-grained perspective to shed light on some important issues related to how to organize NPD through leveraging external sources of innovation. Specifically, we will articulate our discussion around three different and interrelated issues: (a) "projects" as a boundary design variable for NPD; (b) "projects" as the site of "ambidexterity" in NPD processes; (c) "multi-project" management as a tool for optimizing short- and long-term goals in NPD.

3.5 Bringing the "Project" Level of Analysis in the Foreground

3.5.1 "Projects" as a the Site of "Making in Order to Know" in NPD

We have seen above how systems integration has been proposed as a third form of organization beyond carrying out development tasks in-house and outsourcing them ("firm" and "market"). In elaborating their framework, Brusoni et al. (2001) assume that both technological and product dimensions play a key role in affecting the emergence of different organizational forms and consider the presence of a systems integrator firm as the key characteristic of loosely coupled network organizations. In the case of complex products, where interdependencies are difficult to predict and the rate of component technologies is uneven, the systems integrator outsources detailed design and manufacturing to specialized suppliers but simultaneously maintains centralized R&D activities in house.

According to this framework, "R&D projects" are the main site where the firm maintains the knowledge related to the whole product system and at the same time develops new ideas for potential architectural changes. However, "development projects" are the site where NPD performance is generated and where the systems integrator must be able to mobilize its integration competences (see also West and

Iansiti 2003). Much of the knowledge needed at the project level is tacit and very specific, including knowledge on the integration of components into systems and the product model as a whole (Kogut and Zander 1992; Nonaka and Takeuchi 1995; Polanyi 1966). This means that during NPD projects, i.e., when firms' engineers are asked to coordinate suppliers' development activities, if engineers lack this kind of tacit knowledge negative performance consequences would probably arise (cf. Lincoln et al. 1998). Indeed, because of knowledge transfer problems from central R&D departments and development projects, competences held at the firm level do not necessarily turn into actual product development competences at the project level.

From our perspective, one of the most powerful solutions that the firm can adopt to secure the appropriate knowledge required for system integration is exploiting learning by doing opportunities (Zirpoli and Becker 2011a). To effectively capture these opportunities, however, it is required the firm's direct involvement in development work at the project level. Therefore, the firm should not renounce to "making in order to know" at project level. In fact, this mechanism appears crucial for acquiring not only knowledge about technical interdependencies existing among components and subsystems but also knowledge about technologies involved in component development, in other words to ensure that firms "know more that they make", as prescribed by Brusoni et al. (2001). To sum up, both forms of knowledge are necessary to integrate the whole system effectively and to develop competences for dealing with potential architectural changes (Henderson and Clark 1990).

The importance of "making" at the project level points to the necessity of shifting the focus of attention from firms' boundaries and the problem of the boundary definition itself to the development projects, so as to gain a more fine-grained analysis of some important phenomena that take place at the project level. The exposed argument underlines the idea that rather than choosing among two mutually exclusive choices, i.e., outsourcing or developing in house design and engineering tasks, the firm can adapt the proportion with which the same project is carried out in-house and by suppliers concurrently. Our argument resembles (and, thus, is supported by) an important phenomenon empirically documented by recent literature, called "concurrent sourcing" (Parmigiani 2007; Parmigiani and Mitchell 2009): rather than simply choosing between "making" or "buying," the firm chooses to do both simultaneously. What matters about "concurrent sourcing" is the simultaneous use, rather than the extent, of making and buying inputs. The important point is that according to Parmigiani (2007, p. 286), a "small degree of making (or buying) can provide significant benefits" already. Parmigiani's argument raises the possibility that applying this principle in the context of complex product development might hold benefits since it offers the chance of learning by doing while not reneging on benefits related to outsourcing development work.

3.5.2 "Projects" as the Site of "Ambidexterity" in NPD Processes

In effectively organizing new product development through leveraging external sources of innovation, the systems integrator has to manage a crucial tradeoff: on the one hand, it needs to access the specialized knowledge of suppliers; on the other hand, it has to rely on learning by doing mechanisms. The need to pursue such a balance can be framed in terms of the classical dichotomy between exploration and exploitation (March 1991). In fact, when it decides to rely on external suppliers to develop new complex products, the firm can exploit specialized technical knowledge already available somewhere (namely outside the firm's boundaries) and pool them with its current capabilities, thus saving the time and financial resources that would be required to internally develop or maintain skills and competences. Conversely, since developing new products internally offers learning by doing opportunities, this choice puts the firm in the position to expand its knowledge base, acquiring new technical specialized knowledge about specific technologies and components as well as developing new systems integration capabilities.

Since both exploration and exploitation are fundamental for an organization's long-term success (March 1991; Levinthal and March 1993), increasing attention has been given to the issue of "organizational ambidexterity" (Duncan 1976; Tushman and O'Reilly 1996), i.e., the capability of a firm to simultaneously exploit existing competencies and explore new opportunities. Ambidexterity research, however, has mostly focused on the corporate and business unit level of the organization or on the individual level, identifying structural and contextual forms. In "structural ambidexterity" (e.g., Benner and Tushman 2003; O'Reilly and Tushman 2004), a business unit may become ambidextrous through creating two separate functions or subdivisions, respectively, specialized in exploration and exploitation; in "contextual ambidexterity" (e.g., Gibson and Birkinshaw 2004; Mom et al. 2007) ambidexterity is rooted in individuals who have the ability to explore and exploit under the influence of cultural factors (e.g., the context) that orient their behavior toward one or the other.

Also, when the specific context of new product development is taken into account, the ambidexterity discussion remains mainly confined to the organizational level of analysis, while scant attention is given to higher or lower levels (e.g., the project level) in an organization. Perhaps this narrow focus can be understood in light of O'Reilly and Tushman's influential work, who highlight the corporate and business unit level of the organization or, in the context of projects, the individual level (project leaders) (O'Reilly and Tushman 2008; Tushman et al. 2010).

Rather, we argue that in the context of complex product development, since the innovating firm has to leverage external sources of innovation while developing internal competences (thus performing both exploration and exploitation), the project level i.e., a meso-level between the individual and the whole organization, should be seriously considered in investigations of how the firm achieves ambidexterity, which organizational mechanisms enable ambidexterity, and which are the effects of ambidexterity on product development performance. In our argument we draw, in particular, on the still scanty recent research that has started to address the role of projects in providing ambidexterity (e.g. Ahn et al. 2006; Hoang and Rothaermel 2010; Liu and Leitner 2012). Ahn et al. (2006), in particular, were among the first scholars to advance the idea that the project-level can contribute to organizational ambidexterity. Analyzing NPD projects in the information and telecommunications service industry they investigate how high levels of outcomes, in terms of business and knowledge performance at the level of development projects, can be achieved when exploration is combined with exploitation at project level. Building on the idea that internal R&D experience is critical to build absorptive capacity (Cohen and Levinthal 1990), Hoang and Rothaermel (2010) focus on the project level of analysis to identify how exploration and exploitation activities are interrelated in the new product development process, and in particular how the firm draws on existing internal experience to leverage external sources of innovation. As pointed out by the authors (2010:753), taking projects as units of analysis enables "more nuances [to] emerge because we are able to study each individual project rather than outcomes at the more aggregate and theoretically distant firm level." More recently, also Liu and Leiter (2012) have called for specifically considering ambidexterity in projects. Arguing that theirs "is the first study to have identified a link between project team ambidexterity and project team performance" (Liu and Leitner 2012, p. 206), the authors point to the exclusive focus of prior research on organizational-level ambidexterity and highlight the importance of conducting further research on how ambidexterity can be achieved at the project level and the effects that it produces on project performance.

3.6 Conclusion

From the previous discussion, it clearly emerged that in organizing complex new product development, firms cannot completely relinquish learning by doing opportunities. Actually, the positive effects deriving from organizational learning in terms of sustained innovation capability become visible mainly in the long term, since it is through repeatedly drawing on learning by doing contexts that the firm is put in the position to develop new skills and competences, to build its absorptive capacity and adapt to radical technological changes.

In our argument, we pointed to the crucial role played by development projects as the site where these learning opportunities should be exploited by the systems integrator firm (see in particular Sect. 3.1). At the same time, learning in single projects is clearly not sufficient to assure that firm's competences cumulate over time: knowledge transfer across different development projects is also required for achieving the above long-term benefits. However, in applying the project lens to understand how learning by doing is achieved in new product development and how it affects innovation performance and firm's success in the long term, we cannot ignore the empirical results that the extant literature focused on projects has hitherto documented with regard to the role of projects as an organizational form for learning and knowledge transfer. This research, in fact, has documented how "learning is not a natural outcome of projects" (Ayas 1997, p. 59) and that a number of barriers exist to learning both within and across projects (e.g. Bresnen et al. 2004; Newell and Edelman 2008; Swan et al. 2010; Williams 2003; von Zedtwitz 2002). Focusing, in particular, on inter-project learning, Prencipe and Tell (2001) have pointed to the importance of adopting specific organizational mechanisms for transferring knowledge acquired in one project to other projects, while Nobeoka (1995) has pointed to the role of inter-project linkages for effective learning among multiple projects.

In light of what emerged from specific research on projects, by choosing to focus on the development project as unit of analysis, future empirical research on new product development might investigate to what extent mechanisms identified in the literature for overcoming barriers to learning and knowledge transfer can be effectively implemented also when the boundaries of the projects do not fall within the firm's boundaries and external suppliers are actively involved in design and engineering tasks, co-innovating with the leading firm. At the same time, zooming in to the project level, new organizational mechanisms might be identified considering in particular the influence that the specific nature of development projects (for example incremental vs. radical) have on the effective adoption of such mechanisms or looking at the effects that multi-project management can produce in terms of accrued capability of capturing learning by projects as well as sharing and diffusing knowledge across projects and to the wider organization.

With specific regard to the last point, in particular, we built a fruitful bridge for classic work on multi-project management that has focused its attention to the specific context of new product development (e.g., Cusumano and Nobeoka 1998; De Maio et al. 1994; Leonard-Barton 1992; Wheelwright and Clark 1992a, b). In these studies the issues under investigation concern project portfolio selection, strategy, and coordination since it is assumed that multi-project management tools affect product development performance. Future research, thus, should consider in more detail how to leverage specific multi-project management solutions in the context of open innovation not only to optimize short-term product development performance (e.g., shorter lead times and lower development costs) but also long-term goals, namely learning in development projects and transfer knowledge among projects.

References

Ahn, J. H., Lee, D. J., Lee, S. Y., (2006). Balancing business performance and knowledge performance of new product development: Lessons from ITS industry. *Long Range Planning* 39(5): 525–542.

Ancona, D. G., Caldwell, D., (1990). Beyond boundary spanning: Managing external dependence in product development teams. *The Journal of High Technology Management Research* 1(2): 119–135.

Argote. L., (1999). *Organizational learning: creating, retaining, and transferring knowledge.* Boston: Kluwer.

Argote, L., Epple, D., (1990). Learning curves in manufacturing, *Science* 247(Feb): 920–924.

Ayas, K., (1997). Integrating corporate learning with project management. *International Journal of Production Economics* 51(1–2): 59–67.
Bahemia, H., Squire, B., (2010). A contingent perspective of open innovation in new product development projects. *International Journal of Innovation Management* 14(4): 603–627.
Baldwin, C. Y., Clark, K. B., (2000). *Design rules: Volume 1. The power of modularity*. Cambridge, MA: MIT Press.
Benner, M. J., Tushman, M. L., (2003). Exploitation, exploration, and process management: the productivity dilemma revisited. *Academy of Management Review* 28(2): 238–256.
Bonesso, S., Comacchio, A., Pizzi, C., (2011). Technology sourcing decisions in exploratory projects. *Technovation* 31(10–11): 573–585.
Bresnen, M., Goussevskaia A., Swan J., (2004). Embedding new management knowledge in project-based organizations. *Organization Studies* 25(9): 1535–1555.
Brusoni, S., Prencipe, A., (2001). Unpacking the black box of modularity: technologies, products and organizations, *Industrial & Corporate Change* 10(1): 179–205.
Brusoni, S., Prencipe, A., (2006). Making design rules: a multidomain perspective. *Organization Science* 17(2): 179–189.
Brusoni, S., Prencipe, A., Salter, A., (1998). Mapping and measuring innovation in project-based firms. *CoPS Working Paper* No. 46, SPRU, University of Sussex.
Brusoni, S., Prencipe, A., Pavitt, K., (2001). Knowledge specialization, organization coupling, and the boundaries of the firm: Why do firms know more than they make? *Administrative Science Quarterly* 46(4): 597–625.
Campagnolo, D., Camuffo, A., (2010). The concept of modularity in management studies: a literature review. *International Journal of Management Reviews* 12, 259–283.
Cassiman, B., Di Guardo, M.C., Valentini, G., (2010). Organizing links with science: Cooperate or contract? A project-level analysis. *Research Policy* 39(7): 882–892.
Chesbrough, H.W., (2003). *Open innovation: the new imperative for creating and profiting from technology*. Boston: Harvard Business School Press.
Chesbrough, H., Kusunoki, K., (2001). The modularity trap: innovation, technology phases shifts and the resulting limits of virtual organisations, in Nonaka, I., Teece D. (eds). *Managing Industrial Knowledge: Creation, Transfer and Utilization*. (pp. 202–230). London: Sage Press.
Christensen, J.F., (2006). Wither core competency for the large corporation in an open innovation world?, in Chesbrough, H., Vanhaverbeke, W., West, J. (eds). *Open Innovation: Researching a New Paradigm*. (pp. 35–61). Oxford: Oxford University Press.
Clark, K.B., (1989). Project scope and project performance: the effect on parts strategy and supplier involvement in product development. *Management Science* 35(10): 1247–1263.
Clark, K.B., Fujimoto T., (1991). *Product development performance. strategy, organization and management in the world auto industry*. Boston, MA: Harvard Business School Press.
Cohen, W.M., Levinthal, D.A., (1990). Absorptive capacity: a new perspective on learning and innovation. *Administrative Science Quarterly* 35(1): 128–152.
Cusumano, M., Nobeoka, K., (1998). *Thinking beyond lean: how multi-project management is transforming product development at Toyota and Other Companies*. New York: Free Press.
Davies, A., Brady, T., (2000). Organisational capabilities and learning in complex product systems: towards repeatable solutions. *Research Policy* 29(7–8): 931–953.
De Maio, A., Verganti, R., Corso, M., (1994). A multi-project management framework for product development. *European Journal of Operational Research* 78(2): 178–191.
Dierickx, I., Cool, K., (1989). Asset stock accumulation and sustainability of competitive advantage. *Management Science* 35(12) 1504–1513.
Duncan, R.B., (1976). The ambidextrous organization: designing dual structures for innovation, In Kilman, R.H., Pondy, L.R., Slevin, D.P., (eds). *The Management of Organization Design* (vol.1, pp. 167–188): North-Holland: New York.
Engwall, M., (2003): No project is an island: linking projects to history and context, *Research Policy* 32(5): 789–808.

Fine, C.H., (1998). *Clockspeed: winning industry control in the age of temporary advantage*. Reading, MA: Perseus Books.
Fine, C.H., Whitney, D.E., (1996). Is the make-buy decision process a core competence?, April (*MIT unpublished manuscript*).
Frigant, V., Talbot, D., (2005). Technological determinism and modularity: lessons from a comparison between aircraft and auto industries in Europe, *Industry & Innovation* 12(3): 337–355.
Gann, D.M., Salter A., (2000). Innovation in project-based, service enhanced firms: the construction of complex products and systems. *Research Policy* 29(9): 55–72.
Gibson, C.B., Birkinshaw, J., (2004). The antecedents, consequences, and mediating role of organizational ambidexterity. *Academy of management Journal* 47(2): 209–226.
Helper, S.R., MacDuffie, J.P., Sabel, C., (2000). Pragmatic collaborations: advancing knowledge while controlling opportunism. *Industrial and Corporate Change* 9(3): 443–488.
Henderson, R.M., Clark, K.B., (1990). Architectural Innovation: the reconfiguration of existing product technologies and the failure of established firms. *Administrative Science Quarterly* 35(1): 9–30.
Hippel, von E., (1988). *The Sources of Innovation*. Oxford: Oxford University Press.
Hoang, H., Rothaermel F.T., (2010). Leveraging internal and external experience: exploration, exploitation and R&D project performance. *Strategic Management Journal* 31(7): 734–758.
Hobday, M., (2000). The project-based organization: an ideal form for managing complex products and systems? *Research Policy* 29(7–8): 871–893.
Hobday, M., Davies A., Prencipe, A., (2005). Systems integration: a core capability of the modern corporation. *Industrial and Corporate Change* 14(6): 1109–1143.
Hsieh K.N., Tidd, J., (2012). Open versus closed new service development: the influences of project novelty. *Technovation* 32(11): 600–608.
Iansiti, M., (1995). Shooting the rapids: managing product development in turbulent environments. *California Management Review* 38(1): 37–58.
Iansiti, M., Clark K.B., (1994). Integration and dynamic capability: evidence from product development in automobiles and mainframe computers, *Industrial and Corporate Change* 3(3): 557–605.
Karlsson, C., Ahlström, P., (1996). The difficult path to lean product development. *Journal of Product Innovation Management* 13(4): 283–295.
Keegan, A., Turner, J.R., (2002). The management of innovation in project-based firms. *Long Range Planning* 35(4): 367–388.
Kessler, E.H., Bierly, P.E., Gopalakrishnan, S., (2000). Internal vs. external learning in new product development: effect on speed, costs and competitive advantage. *R&D Management* 30(3): 213–222.
Kogut, B., Zander, U., (1992). Knowledge of the firm, combinative capabilities, and the replication of technology. *Organization Science* 3(3): 383–397.
Larson, E.W., Gobeli, D.H., (1989). Significance of project management structure on development success. *IEEE Transactions on Engineering Management* 36(2): 119–125.
Leonard-Barton, D., (1992). Core capabilities and core rigidities: a paradox in managing new product development. *Strategic Management Journal* 13(S1): 111–125.
Levinthal, D.A., March J.G., (1993). The myopia of learning, *Strategic Management Journal* 14(S2): 95–112.
Lincoln, J.R., Ahmadjian, C.L., Mason, E., (1998). Organizational learning and purchase-supply relations in japan: hitachi, matsushita and toyota compared, *California Management Review* 40(3): 241–264.
Liu, L., Leitner D., (2012). Simultaneous pursuit of innovation and efficiency in complex engineering projects - A study of the antecedents and impacts of ambidexterity in project teams. *Project Management Journal* 43(6): 97–110.
MacDuffie, J.P., (2013). Modularity-as-Property, Modularization-as-Process, and Modularity-as-Frame: lessons from product architecture initiatives in the global automotive industry. *Global Strategy Journal* 3(1): 1–125.

March, J.G., (1991). Exploration and exploitation in organization learning. *Organization Science* 2(1): 71–87.
Mikkola, J., (2006). Capturing the degree of modularity embedded in product architectures. *Journal of Product Innovation Management* 23(2): 128–146.
Miozzo, M., Grimshaw, D., (2005). Modularity and innovation in knowledge-intensive business services: IT outsourcing in Germany and the UK. *Research Policy* 34(9): 1419–1439.
Mom, T.J., Van Den Bosch, F.A., Volberda, H.W., (2007). Investigating managers' exploration and exploitation activities: the influence of Top-Down, Bottom-Up, and horizontal knowledge inflows. *Journal of Management Studies* 44(6): 910–931.
Nesta, L., Dibiaggio, L., (2003). Technology strategy and knowledge dynamics: the case of biotech. *Industry & Innovation* 10(3): 329–347.
Newell, S., Edelman L.F., (2008). Developing a dynamic project learning and cross-project learning capability: synthesizing two perspectives. *Information Systems Journal* 18(6): 567–591.
Nishiguchi, T., (1994). *Strategic Industrial Sourcing*. New York: Oxford University Press.
Nobeoka, K., (1995). Inter-project learning in new product development. *Academy of Management Proceedings* August: 432–436. doi:10.5465.
Nobeoka, K., Cusumano, M.A., (1997). Multi-project strategy and sales growth: the benefits of rapid design transfer in new product development. *Strategic Management Journal* 18(3): 169–186.
Nonaka, I., Takeuchi, H., (1995). *The knowledge-creating company*. New York: Oxford University Press.
O'Reilly, C.A., Tushman, M.L., (2004). The ambidexterous organization. *Harvard Business Review* (April): 1–9.
O'Reilly, C.A., Tushman M.L., (2008). Ambidexterity as a dynamic capability: resolving the innovator's dilemma. *Research in Organizational Behavior* 28, 185–206.
Orlikowski, W., (2002). Knowing in practice: enabling a collective capability in distributed organizing. *Organization Science* 13(3): 249–273.
Parmigiani, A., (2007). Why do firms both make and buy? An investigation of concurrent sourcing. *Strategic Management Journal* 28(3): 285–311.
Parmigiani, A., Mitchell, W. (2009). Complementarity, capabilities, and the boundaries of the firm: the impact of within-firm and inter-firm expertise on soncurrent sourcing of complementary components. *Strategic Management Journal* 30(10): 1025–1132.
Prencipe, A., (2003). Corporate strategy and systems integration capabilities - managing networks in complex systems industries. In Prencipe, A., Davies, A., Hobday, M. (eds). *The Business of Systems Integration*. (pp. 114–132) Oxford: Oxford University Press.
Prencipe, A., Tell, F., (2001). Inter-project learning: processes and outcomes of knowledge codification in project-based firms. *Research Policy* 30(9): 1373–1394.
Pinto, M.B., Pinto J.K., (1990). Project team communications and cross-functional cooperation in new program development. *Journal of Product Innovation Management* 7 (3): 200–212.
Polanyi, M., (1966). *The Tacit Dimension*. Garden City, NY: Anchor Books.
Powell, W.W., Koput, K.W., Smith-Doerr, L., (1996). Interorganizational collaboration and the locus of innovation: networks of learning in biotechnology. *Administrative Science Quarterly* 41(1): 116–145.
Rothwell, R., (1992). Successful industrial innovation - critical factors for the 1990s. *R&D Management* 22(3): 221–239.
Salge, T.O., Farchi, T., Barrett, M.I., Dopson, S., (2013). When does search openness really matter? A contingency study of health-care innovation projects. *The Journal of Product Innovation Management* 30(4): 659–676.
Salvador, F., (2007). Towards a product system modularity construct: literature review and reconceptualization. *IEEE Transactions on Engineering Management* 54(2): 219–240.
Sanchez, R., Mahoney, J.T., (1996). Modularity, flexibility, and knowledge management in product and organization design. *Strategic Management Journal* 17(1): 63–76.
Shenhar, A., Dvir D., (2007). *Reinventing project management: the diamond approach to successful growth and innovation*. Boston: Harvard Business.

Staudenmeyer, N., Tripsas, M., Tucci, C., (2005). Interfirm modularity and its implications for product development, *Journal of Product Innovation Management* 22(4): 303–321.
Sturgeon, T.J., (2002). Modular production networks: a new american model of industrial organization. *Industrial and Corporate Change* 11(3): 451–496.
Swan, J., Scarbrough, H., Newell, S., (2010). Why don't (or do) organizations learn from projects? *Management Learning* 41(3): 325–344.
Takeishi, A., (2001). Bridging inter- and intra-firm boundaries: management of supplier involvement in automobile product development. *Strategic Management Journal* 22(5): 403–433.
Takeishi, A., (2002). Knowledge partitioning in the inter-firm division of labor: the case of automotive product development. *Organization Science* 13(may-june): 321–338.
Takeishi, A., Fujimoto, T., (2003). Modularization in the car industry: interlinked multiple hierarchies of product, production, and supplier systems. In Prencipe, A., Davies, A. Hobday, M. (eds). *The Business of Systems Integration.* (pp 254–278). Oxford: Oxford University Press.
Tatikonda, M.V., Rosenthal, S.R., (2000). Successful execution of product development projects: balancing firmness and flexibility in the innovation process. *Journal of Operations Management* 18(4): 401–425.
Tushman, M.L., O'Reilly C.A., (1996). Ambidextrous organizations: Managing evolutionary and revolutionary change. *California Management Review* 38(4): 8–30.
Tushman, M., Smith W.K., Woody, R.C., Westerman, G., O'Reilly C.A., (2010). Organizational designs and innovation streams. *Industrial and Corporate Change* 19(5): 1331–1366.
Ulrich, K.T., (1995). The role of product architecture in the manufacturing firm. *Research Policy* 24(3): 419–440.
Weigelt, C., (2009). The impact of outsourcing new technologies on integrative capabilities and performance. *Strategic Management Journal* 30(6): 595–616.
West, J., Iansiti, M., (2003). Experience, experimentation, and the accumulation of knowledge: the evolution of R&D in the semiconductor industry. *Research Policy* 32(5): 809–825.
Wheelwright, S.C., Clark, K.B., (1992a). *Revolutionizing product development.* New York: The Free Press.
Wheelwright, S.C., Clark, K.B., (1992b). Creating project plans to focus product development. *Harvard Business Review* March-April, 70–82.
Williams, T., (2003). Learning from projects. *Journal of the Operational Research Society* 54(5): 443–451.
Winch, G., (1997). Thirty years of project management. What have we learned? *Presented at British Academy of Management.* Aston University.
Winter, S.G., Szulanski, G., (2001). Replication as strategy. *Organization Science* 12(6): 730–743.
Zedtwitz, von M., (2002). Organizational learning through post-project reviews in R&D. *R&D Management* 32(3): 255–268.
Zirpoli, F., Becker, M.C., (2011a). The limits of design and engineering outsourcing: performance integration and the unfullllled promises of modularity. *R&D Management* 41(1): 21–43.
Zirpoli, F., Becker, M.C., (2011b). What happens when you outsource too much. *MIT Sloan Management Review* 52(2): 59–64.

Chapter 4
Analysis of In-licensing Decisions at a Project and Firm-Level: Evidence from the Biopharmaceutical Industry

Giulia Trombini

Abstract The chapter investigates how firms organize license-ties in order to reconcile resource access at a project-level with knowledge accumulation dynamics at corporate level. Specifically, we analyze what guides firms in choosing whether or not to combine the license-project with R&D collaboration, when in-licensing external technologies. We contend that the choice should be analyzed considering two levels of analysis: the project level and the company level. The organization of license-ties is dependent on the features of the underlying licensed technology as well as on the structure of the firm's knowledge base. Specifically, we contend that these two levels of analysis create contingencies with one another and determine how firms blend project-resource access with corporate knowledge accumulation dynamics. At the project-level, results show that the distance of the in-licensed technology from the firm's knowledge base leads firms to support the license-tie with an R&D collaboration. Interestingly, our findings show that the structure of the firm's knowledge base moderates the above-described relationship. A high depth of technological capabilities has a negative effect on the relationship: the firm does not need to support the license-project with an R&D collaboration. When the firm's competences are instead spread across multiple technological domains—high breadth of technological capabilities—results indicate that the license-project is likely to be supported by an R&D collaboration.

4.1 Introduction

In several industries, especially technology-intensive ones, the rising complexity of the technological landscape, the emergence of new technological trajectories and the increasing volatility of product markets (shorter product-life cycle and changing

G. Trombini (✉)
Department of Management, Ca' Foscari University of Venezia, San Giobbe Cannaregio, 873, 30121 Venezia, Italy
e-mail: giulia.trombini@unive.it

S. Bonesso et al. (eds.), *Project-Based Knowledge in Organizing Open Innovation*,
DOI: 10.1007/978-1-4471-6509-5_4,

consumer needs) have led to an increasing division of innovative labor among several organizations. In such an open innovation context, project-based partnerships are a salient feature of a firm's R&D activity (Chesbrough 2003; Powell et al. 1996). As firms are now specialized in different phases of the industry value chain, they engage in project-based alliances and network ties to perform R&D activity (Hagedoorn 2002), exchanging knowledge with external partners and cumulating relevant external competences to their innovative activity. Firms organize their innovative activity according to a set of collaborative projects with other organizations through which knowledge, capabilities, and resources are built up.

Within this context, a natural question arises: if R&D activity is project-based organized, how should firms structure project-based external ties in order to guarantee an effective resource access to external technologies and skills, and concurrently a coherent integration of the acquired resources into the firm's knowledge endowment?

Past literature on project-based organizing has analyzed the distinctive features of project-based organizations (Cattani et al. 2011). Specifically, the stream of literature on projects as temporary organizational configurations within stable firms has clarified the role of projects. They represent coordination mechanisms through which permanent firms address a given knowledge challenge or explore a new knowledge domain (Brady and Davies 2004; Davies et al. 2011; Prencipe and Tell 2001). Although this literature has greatly enhanced our understanding of project-based organizing, the mechanisms through which firms reconcile knowledge access at a project-level with knowledge accumulation at a corporate level have been overlooked. Little is known about the governance mechanisms firms adopt to organize project-ties with external organizations and how they structure these ties in order to guarantee resource access and accumulation at a project level that is coherent with the firm's knowledge endowment.

In this chapter, we aim to tackle this gap in the literature, by shedding light on how firms structure in-licensing activity. We specifically focus our attention on technology licensing since, among the different forms of project-based cooperative relationships, it represents the lion's share (Anand and Khanna 2000; Hagedoorn 2002). In many technology-intensive industries, licensing importance is growing, since it represents an effective and flexible mechanism for firms to access externally developed knowledge and integrate it into the firm's knowledge base (Laursen et al. 2010).

We investigate the governance mechanisms firms adopt to perform in-licensed projects and how firms reconcile resource access at a project-level with knowledge accumulation dynamics at corporate level. In order to do so, we analyze when firms combine the licensed-project with other forms of technological collaboration, namely R&D collaboration. We contend that the choice of project governance mechanisms is dependent on the strategic aim of the in-licensed project and consequently on the features of the underlying licensed technology as well as on the structure of the firm's knowledge base. Indeed, the interaction between the in-licensed technology and the structure of the firm's knowledge base determine how firms blend project-resource access with corporate knowledge accumulation dynamics.

The chapter considers as its research setting the global biopharmaceutical industry, analyzing in-licensing decisions of the 20 largest biopharmaceutical companies over the time span 1985–2004. The research design relies on multiple sources of information. License-project information such as the in-licensed patents, and the mechanisms combined with licensing are matched with firm's data on patenting activity, knowledge structure, and size. Unlike the majority of studies, data on license-projects are collected at a transaction level of analysis. This fine-grained level of analysis combined with the information of firm's overall knowledge base provides the opportunity to investigate how firms structure in-license projects to coherently blend resource access at project-level with knowledge accumulation dynamics at corporate-level.

The remainder of the chapter is organized as follows. First, we briefly review the literature on project-based organizing through licensing. Second, we derive the theoretical framework and research hypotheses. We then describe the research methods and results emerging from the analysis. Finally, we conclude by discussing the empirical evidence emerging from the analysis and the implications for the literature on project-based organizing.

4.2 Theory and Hypotheses

The importance of accessing external knowledge is well established in innovation literature and the relevance of project-based partnerships as vehicles to access and cumulate other firms' skills and resources, and to gain strategic advantage are well documented in the literature (Chesbrough et al. 2006; Hagedoorn 1993, 2002). Project-ties can be used as boundary spanning tools through which firms access resources at project level and try to integrate and recombine them with the in-house knowledge base. Several authors have investigated the strategic benefits firms can derive by entering into project-based ties with other firms (Baum et al. 2000; Laursen and Salter 2006; Mowery et al. 1996; Nooteboom et al. 2007; Rothaermel and Deeds 2004).

Among the several governance forms available to firms to organize project-based ties, licensing represents the most widely used partnership-mode (Anand and Khanna 2000). Due to its flexibility, it represents the fastest and least risky way by means of which firms can perform external-boundary spanning. Recent studies have highlighted its multiple benefits. In-licensing provides access to proven technologies and allows firms to stay at the technological frontier (Atuahene-Gima 1992). Other recent empirical studies have showed the positive effect of in-licensing on firm's innovative activity, highlighting its positive impact both on firm's search strategy as well as time to invention (Laursen et al. 2010; Leone and Reichstein 2012).

Despite the positive benefits related to in-licensing, it may be difficult for firms to integrate the license-project skills and resources into the firm's knowledge base. If on the one side, license-based ties provide fast and flexible access to advanced knowledge and specialized skills, on the other, the accumulation and integration of these project-based resources to the firm's knowledge base is nontrivial. License-based projects, in fact, may involve little interaction between partners, putting at risk the capability of the recipient firm's personnel involved in the project to effectively understand the technology and to recombine it with the firm's knowledge endowment. Indeed, it seems crucial to understand how firms should organize license-based projects in order to avoid these risks and maximize the benefits from license-projects. The structuring of these ties, in fact, should properly balance resource access at project-level with coherent resource integration at company-level.

A way for firms to reconcile the project-level with the company-level is by supporting the license-project with additional governance mechanisms. Indeed, past studies related to the structuring of license-ties, pinpoint how firms frequently combine license-ties with other forms of project-based collaborations (Hagedoorn et al. 2008). The combination of the license with other forms of technological collaboration allows firms to blend resource access via the license-project with resource accumulation and integration into the firm's knowledge base (Trombini and Comacchio 2012).

Following this emerging stream of literature, we focus our attention on how firms organize license-ties and what the implications are of the features of the in-licensed technology and the structure of the firm's knowledge base in this process. We contend that the choice of project governance mechanisms are dependent on the strategic aim of the in-licensed project and consequently on the features of the underlying licensed technology as well as on the structure of the knowledge base of the firm insourcing the technology. Specifically, we maintain that the interaction between the features of the in-licensed technology and the structure of the firm's knowledge base determine how firms blend project-resource access with corporate knowledge accumulation dynamics.

In order to shed light on this, we focus our attention on a specific governance mechanism that firms may combine with the license-project: R&D collaboration. We focus our attention on this specific governance mechanism since, among the different project-based partnership forms, it presents coordination and communication mechanisms rich enough to provide firms with the opportunity to effectively recombine the in-license project into the firm's knowledge base (Hagedoorn and Hesen 2007). R&D collaboration partnerships are project-based collaborations that span a medium-term time period. For instance, in the pharmaceutical industry, they last on average between 4 and 8 years (DiMasi et al. 2003). The parties usually agree to act collaboratively, and share common goals toward the development and commercialization of a specific technology. They can either pool funds for co-developing and co-marketing the technology; or the joining party can buy into the project and finance its subsequent development, relying on its partner's technical competences. Project managers are appointed by both partners and are responsible for interfirm communications and knowledge sharing. Usually, a research committee composed of two or more representatives of each party coordinates the partnership (Hagedoorn and Hesen 2007). During the collaboration, interfirm communication relies mainly on quarterly meetings, sharing of research facilities, extended visits by research personnel. Through its collaboration and communication mechanisms, the R&D collaboration provides the firm with the opportunity to interact with the partner and assimilate knowledge and skills relative to the licensed project (Trombini and Comacchio 2012). It favors interorganizational learning: the licensee can rely on the licensor's skills and know-how to assimilate and integrate the licensed project within its in-house knowledge endowment.

In the next section, building on organizational learning and absorptive capacity literature, we provide theoretical arguments on how firms organize licensed project ties, by deciding whether or not to combine the license-project with R&D collaboration in order to effectively perform boundary spanning at a project-level and coherently integrate the licensed resources into the firm's knowledge base.

4.2.1 Distance of the In-licensed Technology

Past licensing literature pinpoints that license-ties can be used as technological scouting tools both for explorative search to access distant technological domains with respect to firm's core competences (Laursen et al. 2010), as well as exploitative search to access knowledge that is close to the firm's technological endowment (Lowe and Taylor 1998). However, when a firm, through license-ties, wishes to explore new technological resources that are distant from its technological background, it may face difficulties in exploiting this knowledge and recombining it with its internal knowledge endowment. Indeed, organizational learning literature highlights that firm's directions of technological accumulation are strongly influenced by what the firm already knows (Pavitt 1998). Cognitive obstacles, such as organizational routines, shared knowledge and information filters, make it difficult for an organization to assimilate and recombine into its technological endowment, knowledge that is outside its core competencies (Cohen and Levinthal 1990).

In fact, if on the one side the license-tie provides fast and flexible access to the licensed technologies and resources, on the other, it does not entail coordination and communication mechanisms that may favor the licensee's personnel to interact and learn from the partner, easing the integration of the licensed project resources into the firm's overall knowledge endowment. In this circumstance, we expect the firm, insourcing the project, to combine the license-tie with an R&D collaboration. Through the coordination mechanisms of the R&D partnership, the licensee can enjoy the support of its partner, thereby overcoming assimilation and recombination issues (Trombini and Comacchio 2012). By supporting the license-project with an R&D collaboration, the firm can reconcile knowledge access at a project-level with effective knowledge accumulation dynamics at portfolio-level.

Hypothesis 1 The higher the distance of the in-licensed technology from the firm's knowledge base, the more likely it is that the license is combined with an R&D collaboration.

4.2.2 Distance of the In-licensed Technology and Depth of Firm's Knowledge Base

Past organizational learning literature pinpoints that the dimensions along which a firm's knowledge base can be characterized are depth and breadth. Depth of technological capabilities represents the firm's level of expertise within a specific technological field, and more specifically its capability to innovate within a particular technological niche (George et al. 2008; Sampson 2005).

We hypothesize that the depth of firm's knowledge base negatively moderates the relationship between the distance of the in-licensed technology and the choice of supporting the license-project with an R&D collaboration. Two mechanisms underlie our hypothesis. First, depth builds absorptive capacity within an organization and enables it to understand and assimilate new knowledge generated outside the firm (Cohen and Levinthal 1990). In particular, the specialization in specific technological niches provides in-house employee scientists with a focused common knowledge set, according to which they scout the external environment and select the appropriate technology components to produce fruitful innovation (George et al. 2008). Hence, when the licensed project involves a technology that is distant from the firm's technological background, the depth of the knowledge base in another domain provides a common ground in which employee scientists can integrate the license-project distant knowledge and coherently recombine it within the firm's knowledge endowment.

The second mechanism is related to the fact that depth of technological capabilities provides a deep understanding of casual linkages within the niche. It follows that, when in-licensing a distant technology, it is as if the firm has recognized the potential value of the acquired resources relative to the firm's prior knowledge and has the capability to exploit these resources in relation to its specialized technological endowment. Indeed, when the depth of the technological portfolio is high, the firm is interested in accessing the distant technology through the license-project and exploiting this knowledge only in relation to its technological specialization. There is no need to resort to an R&D collaboration to support the license-project, since the firm does not aim to learn from its partner any complementary knowledge, but rather to access this distant knowledge and exploit it in relation to its specialized knowledge base.

Hypothesis 2 For high levels of depth of the firm's technological capability, the higher the distance of the in-licensed technology from the firm's knowledge base, the less likely it is that the license is combined with an R&D collaboration.

4.2.3 Distance of the In-licensed Technology and Breadth of the Firm's Knowledge Base

Breadth of technological capabilities refers to the scope of a firm's knowledge base, namely the set of different technological domains in which the firm operates (George et al. 2008). When a firm's knowledge base is spread across a large number of technological niches, it is likely that knowledge accumulation dynamics are nontrivial. The firm has to absorb knowledge from several sources and commit resources to multiple domains. The accumulation of technological capabilities, however, is a resource-intensive and time-consuming activity (Dierickx and Cool 1989). Hence, when a firm operates in several technological niches, it is unlikely to simultaneously invest in accumulating capabilities within each domain. Under this circumstance, we hypothesize that knowledge breadth positively moderates the relationship between the distance of the in-licensed technology and the choice of supporting the license-project with an R&D collaboration. Given the strategic importance for the firm to access and cumulate resources from heterogeneous domains, it is key for the firm to guarantee a proper integration of the in-licensed technology into the internal knowledge endowment. However, given the scope of knowledge domains, it would be costly and time-consuming for the firm to integrate the knowledge via the license-project itself. Hence, under these conditions, we expect the firm to combine the license with an R&D collaboration agreement. The latter, in fact, would ease and accelerate the resource integration process, by providing the firm with the opportunity to interact and learn from the license partner in absorbing the licensed technology.

Hypothesis 3 For high levels of breadth of firm's technological capability, the higher the distance of the in-licensed technology from the firm's knowledge base is, the more likely it will be that the license is combined with an R&D collaboration.

4.3 Data and Methods

4.3.1 Research Setting

The research context of the study is the global biopharmaceutical industry in the period 1985–2004. The industry represents the ideal research setting to investigate how firm structure in-license ties in order to blend resource access at project-level with knowledge accumulation at company-level. Innovative activity in the industry, in fact, is typically conducted in the form of joint research projects with a wide variety of partners, ranging from universities and governmental agencies to small and large companies (Pisano 1991, 2006). The industry, indeed, presents the highest partnership frequency among R&D-intensive industries.

The research sample of the study focuses on the largest 20 biopharmaceutical company worldwide. Two motives underlie our choice. First, we restrained our attention to large established firms, since they manage large portfolios of concurrent external project-ties (Arora and Gambardella 1990). Due to the dispersion of knowledge among different actors along the value chain, it is key for them to access these heterogeneous resources and to integrate them into their internal knowledge endowment, in order to feed their internal innovative activity and stay at the knowledge-frontier. Second, among project-based ties through which large firms scout the external environment, licensing represents the lion's share (Roijakkers and Hagedoorn 2006).

4.3.2 Data and Measures

The empirical analysis is based on a sample of 186 license projects, signed by the 20 largest biopharmaceutical firms over the time period 1985–2004. Data on the in-licensing activity of the largest biopharmaceutical firms have been collected from ReCap Database. The database provides access to original license contracts and to a detailed set of information relative to the contracts that were signed concurrently with the license, the patents exchanged through the license, the development stage of the licensed technology, the therapeutic area of the deal, the type of investments planned by the licensee, and the deal compensation scheme.[1] The availability of original license contracts and their corresponding analysis made it possible to collect detailed and objective information relative to each in-licensed project.

Once the data relative to the in-licensed projects had been collected and coded, we matched the dataset with other information sources. Specifically, given the aim of the study, we collected patent information at corporate level for each firm in the sample. In order to do so, two primary sources were used: Who Owns Whom database (2000 version), in order to reconstruct the whole corporate structure of the firms[2]; the National Bureau of Economic Research (NBER) dataset (Hall et al. 2001) to obtain the firms' patent portfolios and statistics related to their innovative activity. Further additional information sources were Biospace, and Compustat. We matched the available information with data on the firm's size and on the partner's primary reference market—SIC codes—in order to detect whether license partners were drug or biotech companies.

Dependent variables

Our model investigates the likelihood that the license-project is combined with an R&D collaboration. Hence, the dependent variable is binary: 1 indicating that the license-tie is combined with an R&D collaboration; and 0 otherwise.

Independent variables

Our primary independent variable is the distance of the *i*th in-licensed project from the firm's knowledge base. Through license-ties, the licensee has access to a set of patents that it has the right to use and exploit within the terms of the contract. Following prior literature (Laursen et al. 2010; Ziedonis 2007), we consider the distance of the in-licensed technology as the degree of overlap between the firm's

[1] For further information on the dataset, refer to Trombini and Comacchio (2012).

[2] Since the firms in the sample operate in the markets for technologies through a set of affiliates and subsidiaries, we employed Who Owns Whom database (2000 edition) to reconstruct each firm's company structure in terms of subsidiaries and affiliates. We also took into account major M&A activities during the period under analysis.

knowledge base and the in-licensed patents. Specifically, we measure the distance as follows:

$$D_i = 1 - \frac{\sum_{t=j-6}^{j-1} p_{kt}}{\sum_k \sum_{t=j-6}^{j-1} p_{kt}} \tag{4.1}$$

where p_{kt} denote the licensee's granted patents that were applied for in the tth year in the same kth four-digit IPC class of the in-licensed patents. We consider the sum of the licensee's granted patents that were applied for in the 6 years before the license-project and are in the same four-digit IPC class of the in-licensed patents divided by the sum of all patents that were applied by the licensee in the same time period. The index varies between 0 and 1, with values close to 1 indicating the highest technological distance between the in-licensed technology and the firm's knowledge base.[3]

In order to test how the interaction between the in-licensed technology and the structure of the firm's knowledge base, determines how firms blend project-resource access with corporate knowledge accumulation dynamics, we consider the firm's knowledge base structure. Following prior studies, we employ two well-established metrics to measure the two dimensions of the firm's technological portfolio: breadth and depth.

Knowledge base depth is proxied through a measure of patent concentration across IPC patent classes (four digit level):

$$Knowledge\ base\ Depth = \sum_k \left(\frac{\sum_{t=j-6}^{j-1} p_{kt}}{\sum_k \sum_{t=j-6}^{j-1} p_{kt}} \right)^2 \tag{4.2}$$

The measure captures the degree of specialization of the firm's technological capabilities in specific technological fields. Values close to the maximum value of 1 indicate a high knowledge depth.

Knowledge base Breadth is measured by the sum of IPC classes in which the firm has applied for patents in the previous 6 years to the license:

$$Knowledge\ breadth = \sum_k c_k \tag{4.3}$$

[3] When the license-project involves the transfer of more than one patent, we computed the distance measure for each in-licensed patent and then considered the average distance.

c_k is a dichotomous variable that assumes value 1 if the firm has applied for patents in the kth IPC class in the previous 6 years to the license-project, and 0 otherwise. The higher the number of different technical classes in which the firm has patents is, the broader will be the firm's knowledge base in terms of technological fields.

As mentioned before, we are interested in analyzing the structure of license-ties factoring the features of the underlying licensed technology as well as on the structure of the firm's knowledge base. Specifically, we test how the interaction between the in-licensed technology and the structure of the firm's knowledge base determines how firms blend project-resource access with corporate knowledge accumulation dynamics. In order to account for such effects, we also compute the interaction effects, respectively, between: *the distance of the in-licensed project and the firm's knowledge depth; the distance of the in-licensed project and the firm's knowledge breadth*.

Control variables

In order to account for potentially competing explanations that might affect how firms structure license-ties, we include in the analysis a set of control variables that, according to prior studies, may affect the choice.

Licensee's size is measured by taking into consideration the number of employees of the licensee in the year before the license is signed (we consider a log transformation).

Vertical license is a dummy variable taking value of 1 if the licensor is a biotech company. Past studies highlight that when partners master different sets of disciplines and technologies, and specifically, when the licensor is a biotech company, partners are likely to combine the license with an R&D collaboration in order to ease knowledge transfer relative to the licensed-project.

Year dummies are included in the model in order to take into account industry shocks and dynamics that may affect how firms organize in-licensed projects.

4.3.3 Estimation Strategy

Given the categorical nature of the dependent variable, the empirical analysis is based on the estimation of a discrete choice model. Specifically, considering our focus on in-licensing activity and hence on the licensee's perspective, we estimate a conditional logit model. This model sets out the specific attributes of the focal organization and the intrinsic randomness of the license governance mechanisms' choice, making it possible to correct the standard errors and the estimates according to the within group correlations (Greene 2003). In this respect, the model may be considered superior to a standard logit specification. For robustness check, we also estimate a standard logit model.

Table 4.1 Sample descriptive statistics

	N	Mean	S.d.	Min	Max
License combined with R&D	186	0.33	0.47	0.00	1.00
Distance of the in-licensed technology	186	0.89	0.10	0.60	1.00
Knowledge base depth	186	0.16	0.06	0.05	0.30
Knowledge base breadth	186	106.76	49.85	17.00	218.00
Vertical license	186	0.71	0.45	0.00	1.00
Firm's size	186	11.13	0.73	8.40	12.00
License R&D focus	186	0.79	0.41	0.00	1.00
Time period				1985	2003

Table 4.2 Correlation matrix

	1	2	3	4	5	6	7
1. License combined with R&D	1						
2. Distance of the in-licensed technology	**0.14**	1					
3. Knowledge base depth	−0.01	**−0.26**	1				
4. Knowledge base breadth	−0.08	**0.27**	**−0.68**	1			
5. Vertical license	**0.13**	0.12	0.07	−0.08	1		
6. Firm's size	0.04	**0.27**	**−0.41**	**0.83**	0.00	1	
7. R&D focus of the license	0.13	−0.01	−0.11	−0.09	0.01	−0.08	1

Correlation coefficients in bold are significant at 5 % level

4.4 Results

Tables 4.1 and 4.2, respectively, report the descriptive statistics and the correlation coefficients among variables used in the analysis. We observe a total of 186 licensеties. The average number of license combinations with R&D collaboration is 33 %. The correlation matrix highlights that multicollinearity is not an issue, except in the case of breadth and depth of technological capabilities, since their correlation coefficient is around the "warning level" −0.6. In order to further investigate potential multicollinearity issues, we computed variance inflated factors, by running "artificial"OLS regressions between each independent variable as the "dependent" variable and the remaining independent variables as suggested by Maddala (2000). As all VIF values are below the threshold level of 4 and this indicates that there is no multicollinearity between the variables. Further, in order to avoid these issues, when computing interaction terms, the main explanatory variables—distance of the in-licensed technology, knowledge base depth, and knowledge base breadth were mean-centered (Jaccard et al. 1990).

Table 4.3 depicts the estimations of the conditional model analysis. Specifically, Model 1 is the control model and is significant ($p < 0.001$). From Model 2 to Model 6, we introduce step-wise the main explanatory variables. For clarity in the discussion of the findings, we will focus on the evidence emerging from Models 4, 5, and 6.

Table 4.3 Conditional logit estimates: likelihood of combining the license with an R&D collaboration

	Model 1	Model 2	Model 3	Model 4	Model 5	Model 6
Technological distance		**4.98***	**5.49***	**6.88***	**8.24***	**8.25***
		(1.310)	(1.586)	(1.700)	(2.024)	(2.014)
Knowledge base depth			−7.01	−8.21	−9.82	−9.82
			(6.573)	(6.429)	(6.822)	(6.835)
Knowledge base breadth			**−0.06***	**−0.07***	**−0.07***	**−0.07***
			(0.017)	(0.018)	(0.020)	(0.020)
Tech. distance × breadth				0.05****		0.00
				(0.033)		(0.044)
Tech. distance × depth					**−57.62****	−56.74****
					(1.840)	(2.949)
Vertical tie	**0.82*****	0.78****	0.70	0.68	0.67	0.66
	(0.391)	(0.405)	(0.438)	(0.440)	(0.447)	(0.448)
R&D focus of the tie	1.40****	1.49****	1.36	1.40	1.39	1.39
	(0.843)	(0.820)	(0.852)	(0.857)	(0.862)	(0.853)
Firm's size	0.67	0.46	1.04	0.98	1.02	1.02
	(0.763)	(0.741)	(0.819)	(0.869)	(0.878)	(0.891)
Year dummies distance	Yes	Yes	Yes	Yes	Yes	Yes
No. observation	186	186	186	186	186	186
Log-likelihood	−78.54	−75.51	−67.95	−67.29	−66.43	−66.43

Robust standard errors in parentheses
* $p < 0.001$, ** $p < 0.01$, *** $p < 0.05$, **** $p < 0.1$

Table 4.3 depicts the estimations of the conditional model analysis. Specifically, Model 1 is the control model and is significant ($p < 0.001$). From Model 2 to Model 6 we introduce step-wise main explanatory variables. For clarity in the discussion of the findings, we will focus on the evidence emerging from Models 4, 5, and 6.

Hypothesis 1 suggests that *the higher the distance of the in-licensed technology from the firm's knowledge base, the more likely it is that the license is combined with an R&D collaboration.* Table 4.3 contains significant positive estimates with regard to the distance of the in-licensed technology with respect to the firm's knowledge base, hence lending strong support to our hypothesis.

Hypothesis 2 is confirmed by the analysis. Models 5 and 6 present significant coefficients. However, since the conditional logit and the logit model are nonlinear models, we followed the directions of Hoetker (2007) to compute the significance of the interaction term: *for high levels of depth of the firm's technological capability, the higher the distance of the in-licensed technology from the firm's knowledge base, the less likely it is that the license is combined with an R&D collaboration.*

Finally, Hypothesis 3 asserts that *for high levels of breadth of the firm's technological capability, the higher the distance of the in-licensed technology from the firm's knowledge base, the more likely it is that the license is combined with an R&D collaboration.* Using the same procedure adopted for testing Hypothesis 2, we find partial support for Hypothesis 3. The coefficient is significant as expected only in

Model 4; however, in Model 6, when introducing also the interaction term between distance of the licensed technology and knowledge base depth, the coefficient is no longer significant.

We run a series of robustness checks to validate the consistency of the results. Specifically, we changed the operationalization of key theoretical variables. For each explanatory variable—technological distance, knowledge base breadth and depth—we estimated their value for different time lags. Instead of considering a 6-year time window before the signature of the license-partnership to classify the firm's patent activity, we considered other time spans (3-, 4- and 5-year time windows). Under the different specifications, results were stable.

As for the variable *depth of knowledge base*, following prior studies, we took into account an alternative measure by calculating the maximum number of patents in any one IPC patent class (four digit level). One disadvantage of the concentration measure, in fact, is that it penalizes firms for dispersion across patent classes (George et al. 2008). Under this alternative specification, results were confirmed.

Finally, relative to the variable *knowledge base breadth*, we normalized the variable in the range (0, 1). Under this different variable specification, results were stable.

4.5 Discussion and Conclusion

The aim of this chapter was to shed light on how firms organize license-based ties and, more specifically, the conditions when the license-project is combined with R&D collaboration. Our focus was to highlight how in-sourcing decisions through licensing are determined both by the project's features—in our case the distance of the in-licensed technology from the firm's technology base—as well as by the firm's knowledge base attributes—the breadth and depth of technological capabilities. In particular, we contend that these two levels of analysis create contingencies with one another.

The empirical analysis shows that the distance of the in-licensed technology from the firm's knowledge base is an important determinant, leading firms to support the license-tie with an R&D collaboration. Due to the diversity of the licensed-project from the firm's core competences, both resource access and knowledge recombination are at risk. Indeed, firms use the R&D collaboration in order to overcome absorptive capacity issues and effectively recombine the in-licensed technology and skills into the firm's knowledge base.

Interestingly, our findings show that the structure of the firm's knowledge base moderates the above-described relationship. More specifically, high depth of technological capabilities has a negative effect on the relationship. A high concentration of the firm's technological competences within specific domains, implies that, when in-licensing a distant technology, the firm exploits it in relation to its focused knowledge endowment. The firm does not need to support the license-project with an R&D collaboration, since it does not aim to gain any complementary knowledge from its partner, but rather to access relevant distant knowledge and exploit it in relation to is specialized knowledge base.

When the firm's competences are instead spread across multiple technological domains—high breadth of technological capabilities—results indicate that the license-tie is likely to be supported by an R&D collaboration. Given the strategic importance of heterogeneous bodies of knowledge for the firm's internal competence accumulation dynamics, it is key for the firm to guarantee an effective integration of the in-licensed distant technology into the firm's technology background. Indeed, the R&D collaboration serves this role, by allowing the firm to interact and learn from the partner, easing and speeding up the knowledge recombination process, which through the license-project itself might be at risk.

Our study extends and contributes to the current academic debate in the following directions. First, we contribute to the broad literature on open innovation (Bianchi et al. 2011; Chesbrough 2003), and, more specifically, on the demand-side of the phenomenon. We throw light on how firms organize external ties so as to combine resource access at project-level with coherent knowledge accumulation dynamics at corporate level. By taking into account both firm's level attributes—the knowledge structure—as well as project-based features—the distance of the in-licensed technology—we provide a detailed analysis of what drives firms when organizing external ties to access and cumulate external knowledge and skills.

Second, the study contributes to the literature on project-based organizing within permanent organizations (Davies et al. 2011). Past literature has clarified the role of project-ties with external partners as coordination mechanisms through which permanent firms address a given knowledge challenge or explore a new knowledge domain. Although this literature has greatly enhanced our understanding of project-based organizing, the mechanisms through which firms reconcile knowledge access at a project-level with knowledge accumulation at a corporate level have been overlooked.

The study, by taking into account both the project-level and the firm-level, provides a detailed analysis of how firms structure external ties in order to guarantee resource access and accumulation at a project-level that is coherent with the firm's knowledge structure, either along a knowledge breadth dimension or depth one.

Thirdly, we contribute to the licensing literature (Anand and Khanna 2000; Hagedoorn et al. 2008). The chapter clarifies how firms use license projects as a means to capture distant technological opportunities outside the boundaries of the firm.

References

Anand, B., Khanna, T. (2000). The structure of licensing contracts. *Journal of Industrial Economics* 48(1): 103–134.

Arora, A., Gambardella, A. (1990). Complementarity and external linkages: the strategies of the large firms in biotechnology. *Journal of Industrial Economics* 38(4): 361–379.

Atuahene-Gima, K. (1992). Inward technology licensing as an alternative to internal R&D in new product development: a conceptual framework. *Journal of Product Innovation Management* 9(2): 156–167.

Baum, J.A.C., Calabrese, T., Silverman, B.S. (2000). Don't go it alone: alliance network composition and startups' performance in Canadian biotechnology. *Strategic Management Journal* 21(3): 267–294.

Bianchi, M., Cavaliere, A., Chiaroni, D., Frattini, F., Chiesa, V. (2011). Organisational modes for open innovation in the bio-pharmaceutical industry: an exploratory analysis. *Technovation* 31(1): 22–33.

Brady, T., Davies, A. (2004). Building project capabilities: from exploratory to exploitative learning. *Organization Studies* 25(9): 1601–1621.

Cattani, G., Ferriani, S., Frederiksen, L., Taube, F. (2011). *Project-based organizing and strategic management*. Howard House, Wagon Lake, UK: Emerald Group Publishing.

Chesbrough, H. (2003). *Open innovation: the new imperative for creating and profiting from technology*. Boston: Harvard Business School Press.

Chesbrough, H., Vanhaverbeke, W., West, J. (2006). *Open innovation: researching a new paradigm*. Oxford: Oxford University Press.

Cohen, W.M., Levinthal, D. (1990). Absorptive capacity: a new perspective on learning and innovation. *Administrative Science Quarterly* 35(1): 128–152.

Davies, A., Brady, T., Prencipe, A., Hobday, M. (2011). Innovation in complex products and systems: implications for project-based organizing. In Cattani, G., Ferriani S., Frederiksen L., Taube, F. (eds). *Project-based organizing and strategic management*. Howard House, Wagon Lane, UK: Emerald Group Publishing Limited.

Dierickx, I., Cool, K. (1989). Asset stock accumulation and sustainability of competitive advantage. *Management Science* 35(12): 1504–1513.

DiMasi, J.A., Hansen, R.W., Grabowski, H.G. (2003). The price of innovation: new estimates of drug development costs. *Journal of Health Economics* 22(2): 151–185.

George, G., Kotha, R., Zheng, Y. (2008). Entry into insular domains: a longitudinal study of knowledge structuration and innovation in biotechnology firms. *Journal of Management Studies* 45(8): 1448–1474.

Greene, W. (2003). *Econometric Analysis*. Upper Saddle River, New Jersey: Prentice Hall.

Hagedoorn, J. (1993). Understanding the rationale of strategic technology partnering: interorganizational modes of cooperation and sectoral differences. *Strategic Management Journal* 14(5): 371–395.

Hagedoorn, J. (2002). Inter-firm R&D partnerships: an overview of major trends and patterns since 1960. *Research Policy* 31(4): 477–492.

Hagedoorn, J., Hesen, G. (2007). Contract law and the governance of inter-firm partnerships: an analysis of different forms of partnering and their contractual implications. *Journal of Management Studies* 44(3): 342–365.

Hagedoorn, J., Lorenz-Orlean, S., van Kranenburg, H. (2008). Inter-firm technology transfer: partnership-embedded licensing or standard licensing agreements? *Industrial and Corporate Change* 18(3): 529–550.

Hall, B., Jaffe, A., Trajtenberg, M. (2001). The NBER patent citations data file: lessons, insights, and methodological tools. *NBER Working Paper* 8498.

Hoetker, G. (2007). The use of logit and probit models in strategic management research: critical issues. *Strategic Management Journal* 28(4): 331–343.

Jaccard, J., Turrisi, R., Wan, C. (1990). *Interaction effects in multiple regression*. Newbury Park, California: Sage Publications Inc.

Laursen, K., Salter, A. (2006). Open for innovation: the role of openness in explaining innovation performance among U.K. manufacturing firms. *Strategic Management Journal* 27(2): 131–150.

Laursen, K., Leone, M. I., Torrisi, S. (2010). Technological exploration through licensing: new insights from the licensee's point of view. *Industrial and Corporate Change* 19(3): 871–897.

Leone, M.I., Reichstein, T. (2012). Licensing-in fosters rapid invention! The effect of the grant-back clause and technological unfamiliarity. *Strategic Management Journal* 33(8): 965–985.
Lowe, J., Taylor, P. (1998). R&D and technology purchase through licence agreements: complementary strategies and complementary assets. *R&D Management* 28(4): 263–278.
Maddala, G.S. (2000). *Introduction to econometrics*. Upper Saddle River, NJ: Prentice-Hall.
Mowery, D.C., Oxley, J.E., Silverman, B.S. (1996). Strategic alliances and interfirm knowledge transfer. *Strategic Management Journal* 17(2): 77–91.
Nooteboom, B., Vanhaverbeke, W., Duysters, G., Gilsing, V., Vandenoord, A. (2007). Optimal cognitive distance and absorptive capacity. *Research Policy* 36(7): 1016–1034.
Pavitt, P. (1998). Technologies, products and organization in the innovating firm: what Adam Smith tells us and Joseph Schumpeter doesn't. *Industrial and Corporate Change* 7(3): 433–452.
Pisano, G. (1991). The governance of innovation: vertical integration and collaborative arrangements in the biotechnology industry. *Research Policy* 20(3): 237–249.
Pisano, G. (2006). Can science be a business? Lessons from biotech. *Harvard Business Review* 84(10): 114–125.
Powell, W.W., Koput, K.W., Smith-Doerr, L. (1996). Interorganizational collaboration and the locus of innovation: networks of learning in biotechnology. *Administrative Science Quarterly* 41 (1): 116–145.
Prencipe, A., Tell, F. (2001). Inter-project learning: Processes and outcomes of knowledge codification in project-based firms. *Research Policy* 30(9): 1373–1394.
Roijakkers, N., Hagedoorn, J. (2006). Inter-firm R&D partnering in pharmaceutical biotechnology since 1975: trends, patterns, and networks. *Research Policy* 35(3): 431–446.
Rothaermel, F.T., Deeds, D.L. (2004). Exploration and exploitation alliances in biotechnology: a system of new product development. *Strategic Management Journal* 25(3): 201–221.
Sampson, R. (2005). Experience effects and collaborative returns in R&D alliances. *Strategic Management Journal* 26(11): 1009–1031.
Trombini, G., Comacchio, A. (2012). Cooperative markets for ideas: when does licensing combine with R&D partnerships? *Working Paper Series, Dipartimento di Management, Università Ca Foscari Venezia* n. 8/2012, Luglio 2012, 1–45.
Ziedonis, A.A. (2007). Real options in technology licensing. *Management Science* 53(10): 1618–1633.

Chapter 5
Open Innovation at Project Level: Key Issues and Future Research Agenda

Sara Bonesso, Anna Comacchio and Claudio Pizzi

Abstract This chapter addresses some key open issues of a project-based approach to open innovation, drawing on the empirical findings and theoretical discussion of the previous chapters. After a brief discussion of the theoretical gaps in the previous literature, we provide arguments in support of the adoption of a project-level of analysis when studying how firms organize open and distributed innovation processes. The chapter tackles two main themes of a project-based approach, identifying fertile avenues for future research. First, it analyzes factors explaining why firms decide to open their boundaries and to organize in-house tasks and outside source activities on a project basis. Antecedents related to the knowledge features of a project are discussed. Second, the chapter draws the attention to the still under-investigated relationship between the project level and the firm level of analysis. It discusses the challenge firms face in managing effectively and efficiently product development projects across boundaries in the short term as well as in building organizational capabilities and knowledge at firm level in the long run.

S. Bonesso (✉) · A. Comacchio
Department of Management, Ca' Foscari University of Venezia, San Giobbe Cannaregio, 873, 30121 Venezia, Italy
e-mail: bonesso@unive.it

A. Comacchio
e-mail: comacchio@unive.it

C. Pizzi
Department of Economics, Ca' Foscari University of Venezia, San Giobbe Cannaregio, 873, 30121 Venezia, Italy
e-mail: pizzic@unive.it

S. Bonesso et al. (eds.), *Project-Based Knowledge in Organizing Open Innovation*, DOI: 10.1007/978-1-4471-6509-5_5,

5.1 Introduction

The book proposes a project-based view on open innovation in contexts where knowledge is dispersed and the locus of innovation resides in a network of specialized external knowledge sources (Powell et al. 1996). We depart from the previous innovation management literature that, notwithstanding a consensus on the fact that innovation strategies are implemented through multiple NPD projects (Gemünden et al. 2013; Wheelwright and Clark 1992), has favored a firm level of analysis in explaining how firms allocate innovative labor within and across boundaries.

We suggest that the aim of a project is not only the development of a new product but also the exploration of external sources of new knowledge in order to find opportunities that can significantly foster a firm's capacity to be innovative (Lenfle 2008; Söderlund et al. 2008). The centrality of the project in a firm's knowledge base development has been shown by studies on complex products and systems (design, engineering and construction), where the project is the site for innovative labor division and knowledge accumulation. A project is conceived as a coordination mechanism to integrate the specialist knowledge and competences of the network of organizations contributing to the NPD process (Brusoni et al. 1998; Brusoni and Prencipe 2006; Prencipe 2003; Zirpoli and Becker 2011). More recently, the spread of a project-based approach in different sectors and firms (Davies and Brady 2000; Hobday 2000) has moved the project to the center of the research agenda of strategy and organization studies and has expanded the empirical investigation from manufacturing to the service sector and from permanent to temporary organizations (Cattani et al. 2011) However, this research stream has provided little empirical evidence on a project-by-project open innovation strategy.

A renewed attention to projects in the context of distributed innovation processes has recently been advanced by the literature on open innovation. This literature singled out the project stages, identifying them as different opportunities for opening up organizational boundaries (Grönlund et al. 2010), and explained which specific portion of the external technological environment a firm might exploit at each project stage. However, little attention has been devoted to the different composition of a firm's project portfolio and how the different features of each project influence the choice of open innovation forms (Christiansen et al. 2013; Huizing 2011). Only recently, the adoption of a contingent approach to innovation has led authors to argue that a fine-grained investigation at the micro level of a project is needed for a deeper understanding of the adoption of open innovation forms and of how firms deal with a wide range of decisions regarding external parties (Bahemia and Squire 2010; Bonesso et al. 2011; Cassiman et al. 2010; Hoang and Rothaermel 2010; Hsieh and Tidd 2012; Salge et al. 2013; West et al. 2014).

Finally, the literature on project portfolios has been primarily focused on the strategic management and the optimization of in-house multiple-projects addressing issues such as strategy formation at the project level and alignment with the firm's strategic goals (MacCormack et al. 2012; Vuori et al. 2013); resource allocation and project portfolio optimization (Laslo 2010; Perks 2007); project management practices and performance (Besner and Hobbs 2013; Cooper et al. 1999); inter-project learning (Brady and Davies 2004; Jerry 2008). Therefore, the previous literature has underestimated the fact that project boundaries and organizational boundaries do not always coincide and firms might leverage on projects to explore new boundary options.

Coherently with recent research on open innovation forms and to overcome the literature gaps, we propose that the project level of analysis provides a useful lens for a fine-grained investigation of how firms strategically choose to jointly explore new product options and organizational boundary options.

The book addresses the question of whether and how a set of discrete and diverse NPD projects is related to a company's choices to open up organizational boundaries and arrange innovative labor among inside units and outside partners. The empirical findings and theoretical arguments provided in previous chapters support the claim for a project-based view of open and distributed innovation. By showing the centrality of organizational choices at the micro level of analysis, the book also offers an insightful theoretical and empirical contribution to the literature on knowledge and innovation management. In the following sections the emerging topics previously highlighted will be discussed along with some quantitative approaches for further statistical investigation.

5.2 Open Innovation Project-by-Project

Do firms make inbound sourcing decisions project-by-project or through a common strategy across the project portfolio? The book raises this research question and provides arguments and empirical evidence on the increasing diffusion and the emerging challenges of a project-by-project open innovation strategy, in those permanent organizations in which innovation is typically organized in the form of projects.

First, the book shows that a project-by-project approach is spreading among firms belonging to different sectors from high-tech to medium-tech settings, coherently with the diffusion of firms adopting a project-based form (Cattani et al. 2011; Hobday 2000). As discussed in Chap. 1, this diffusion is related, on the one hand, to the need to pursue greater flexibility in order to quickly reorient innovation processes according to the market and technological evolution; on the other hand, it is explained by the need to widen the scope of sourcing choices to get access to complementary resources while retaining organizational specialization. Future research should investigate the diffusion of a project-by-project approach in organizing distributed innovation processes in other industrial settings, such as low-tech sectors and in the service sector, which is still investigated at firm level of analysis (Mina et al. 2013).

A further contribution regards the differentiated approaches adopted by firms in designing the network of external sources. Indeed, despite the benefits of the flexibility allowed by a project-based approach, other factors or possible drawbacks of deciding project-by-project may spur some firms to choose an overall firm-based approach to external collaborations. Indeed, findings show that firms diverge according to the degree of flexibility with which they adopt an open innovation approach and design the network of external sources: while some firms tend to implement a prevalent strategy (closed or open) across the whole project portfolio, others adopt a differentiated approach on a project basis. In line with the empirical findings, the book suggests a procedure to study this phenomenon and to distinguish between firms that implement a common inbound strategy in every project and those that reconfigure sources and partners project-by-project. As illustrated in Chap. 2, leave-k-out testing (Bruce and Martin 1989) represents an effective procedure to empirically investigate the degree to which a firm decides to adopt a project-by-project-based approach in allocating innovative tasks across boundaries. From a different point of view, the leave-k-out testing procedure may be seen as a naive cluster analysis (Du 2010) useful in investigating sourcing strategies in small sample size. If this peculiarity is lacking, some clustering techniques linked to the Artificial Neural Network (ANN) might be useful.

Third, proposing a project-based view on open innovation, this volume sheds new light on the main project-based antecedents that spur firms to strategically reconfigure the network of sources and distribute the innovative labor among external partners. Adopting a knowledge-based perspective, the empirical chapters demonstrate how firms flexibly recombine internal and external sources according to the specific knowledge features that characterize each single NPD project. In this regard, the radicalness of the project emerged as a relevant knowledge attribute in explaining open and distributed innovation across different industrial settings. In the case of the machine tool industry, and in general in B2B industries, where new industrial products are developed through projects tailored to customers' needs, novelty is expressed by the degree through which an NPD project meets ahead-of-market needs. The search for original product features and functionalities benefits from the collaboration with external sources, which provides firms with new product components and solutions and enables divergent thinking for the generation of new state-of-the-art product concepts. Moreover, Chap. 2 shows that nonlocal partners seem to enable the identification of cross-cultural differences and diverse customer mindsets. In the biopharmaceutical industry, the inbound decisions are made project by project since each single NPD project may imply a specific in-licensing strategy. Specifically, Chap. 4 investigates two of them, namely stand-alone licensing agreements or licensing agreements combined with R&D collaboration. Research findings show that when a licensee engages in a license-project, in which the distance of the in-licensed technology from its knowledge base is high, the firm gains access to this radical knowledge by combining the license with an R&D collaboration to overcome cognitive barriers that might prevent knowledge learning and absorption.

Another crucial project attribute that emerged as a relevant driver in explaining project-by-project inbound decisions is knowledge breadth, as empirical evidence from the machine tool industry suggests (Chap. 2). This dimension is sound especially for those industrial systems characterized by technological convergence.

Future studies should extend the investigation of knowledge breath at project level to other sectors that are facing a process of hybridization across different technological domains.

These two main knowledge dimensions of a project, namely knowledge novelty and knowledge breadth, could be statistically investigated by using the gravity model (Kimura and Lee 2006; Picci 2010). This model is a possible candidate for studying the inbound decisions taken by a firm, considering knowledge novelty and breadth as measures of the distance of a new project from the firm's knowledge base. The operationalization of the "knowledge distance" could be enriched by considering other approaches, such as the psychic distance adopted by the literature on internationalization (Brewer 2007; Dow and Karunaratna 2006). The psychic distance is conceived as the distance resulting from managerial perceptions of both cultural and business differences (Evans and Mavondo 2002). For instance, Holzmuller and Kasper (1990) measure psychic distance, as perceived by an individual, by using cognitive mapping.

Moreover, this volume shows how the knowledge features at project level impact on different boundary choices. First, project knowledge features affect the decision to search beyond organizational boundaries for novel ideas and solutions. Second, they impact on the forms of governance of the relationship with external partners. Third, the empirical evidence demonstrates that these dimensions, knowledge novelty in particular, spur firms to search cognitive distant sources rather than similar ones.

Finally, the various chapters show that the decision to strategically organize open innovation project by project may not only be motivated by the search for novel knowledge and to tap diverse technological fields, but also by the intention to gain access to the partner's specific technological and organizational capabilities necessary for the development of the NPD projects.

5.3 Reconciling a Project-by-Project Approach with the Firm's Level of Analysis

When firms simultaneously adopt multiple open innovation options on a project basis, they leverage on the knowledge exploration opportunities and the benefits of project portfolio flexibility in the short term. However, a firm managing open innovation from a single project perspective might incur transaction costs and risks (Salge et al. 2013), related to a differentiated external network of partners and to its changing composition over time. Costs and risks can be mitigated by an integrated project-firm-level approach (Faems et al. 2005). Furthermore, a multiple-project approach to open innovation helps to consider the interdependences among projects and the relationship between a project's knowledge-developing aims and the knowledge accumulation dynamics at the company level. Despite the insights provided by the literature that addresses the issue of organizing open innovation at the firm or at the project level (Lichtenthaler 2011), research has mainly remained silent on the possible integration between the two levels.

In line with these considerations, while proposing a project-based perspective for studying in-depth how firms organize innovative labor across boundaries, this volume also suggests the need to reconcile the project level and the firm level of analysis.

As discussed in Chap. 1, as far as the level of a firm's absorptive capacity is concerned (Cohen and Levinthal 1990), the breadth and depth of a firm's knowledge stock (Reich et al. 2012) might affect the firm's propensity to rely on external sources. Consequently, we suggest these firm-level characteristics might interact with the project's knowledge features in explaining the boundary options adopted by a firm. Specifically, a firm's knowledge base could interact with the project-level learning process in two different ways. First, it could affect the decision of leveraging on a specific partner and of choosing a governance mode to access new knowledge. Second, it could impact on how effectively a firm could assimilate external knowledge at organizational level to sustain the long-term competitive advantage.

Projects play a twofold function as regards boundary decisions: they are boundary spanning tools by which firms search and recognize external innovation opportunities and they are the site where a significant learning-by-doing process with external partners takes place. For this reason, when a firm looks outside for new solutions by means of an NPD project, the structure of the firm's knowledge base might mediate the impact of the features of a project in determining which type of partners and governance modes are to be deployed. Moreover, when a firm relies on close cooperation with partners and develops tacit specialized knowledge through learning-by-doing, mechanisms such as transactive memory systems, information pooling, and functional diversity (Gardner et al. 2012) are needed to effectively support the knowledge transfer across the organization and favor its assimilation at company level.

Chapters 1 and 3 theoretically address these issues while Chap. 4 provides explorative evidence on how firms, by adopting a flexible project-by-project approach, are spurred in their boundary choices by the interaction among project features and knowledge-based attributes.

Chapter 4 specifically focuses attention on how the interaction between the features of a technology in-licensed at project level and the structure of the firm's knowledge base affect boundary choices. It shows that the different composition of a firm's technological capabilities (knowledge breadth vs. knowledge depth) moderates the impact that the distance of the in-licensed technology from the firm's knowledge base has on the license-tie choice. Findings from the global biopharmaceutical industry highlight that when the knowledge base is characterized by high depth, a distant partner's resources are not necessary to complement the firm's highly specialized capabilities. Thus, the licensee does not need to leverage on learning-by-doing with the partner by combining license with an R&D collaboration. When a firm's knowledge base is highly spread across multiple technological domains (high breadth) the in-licensed project should be supported by an R&D collaboration. The firm is interested in leveraging the partner's complementary skills and activate a learning process aimed at further broadening the firm's capability set.

The relevance of the learning process at the project level is discussed in Chap. 3; it suggests the usefulness of a project-based approach to study firms operating in complex product industries as a powerful lens through which to understand how firms exploit opportunities to build the company knowledge base by a network of external partners. More specifically, the chapter shows how projects can be strategically managed by the innovating firm to build both component and architectural knowledge and to simultaneously pursue the benefits of exploring and exploiting external sources of knowledge and capabilities (Parmigiani 2007; Parmigiani and Mitchell 2009). Indeed, the project level of analysis provides a fine-grained lens to address the issue of how systems integrators build their capabilities, developing tacit knowledge on components and their interdependences in the long term and how, accordingly, they formulate their make or buy strategy.

Drawing on these explorative insights we propose that future research should investigate more in-depth the processes and the mechanisms by which firms reconcile learning and knowledge exploration at the micro level with knowledge accumulation at the macro level. Moreover, studies should delve into the interaction between the two levels in affecting innovative performance. A model that can consider simultaneously two different levels of analysis, project and firm, is the threshold model (Dagenais 1969, 1975) in which a variable at the firm level, for instance the dimension or the age of the firms, plays the role of switching variable. When this variable exceeds a certain threshold, the parameters of the model at project level present some values that change when the variable decreases until it goes below the threshold.

Finally, the project level should be considered by future studies on organizational mechanisms that enable a company's ambidexterity. This is a promising line of research, considering that the literature has only recently addressed the issue of how ambidexterity can be managed in NPD project teams, as a meso-level between the individual and the whole organization, and what ambidextrous practices can be adopted at this specific level (Liu and Leitner 2012; Turner et al. 2013). For instance, the relationship between ambidexterity and NPD projects across boundaries has been highlighted by Bahemia and Squire (2010), who defined as ambidextrous open innovation projects those that include both new and existing external partners. An interesting avenue for further research would be to investigate how firms, on a project basis, decide to involve different types of actors, exploring new partners' resources, and exploiting long-lasting ones.

5.4 Conclusions and Implications

This volume aims to contribute to the debate on open and distributed innovation, by bridging the recent debate on project portfolio management, project-based firms, and temporary organizations on the one hand (Cattani et al. 2011) and research on innovation management and new open organization forms on the other (Gulati et al. 2012; Lakhani and Tushman 2012). We claim that a project-based perspective in analyzing the division of innovative labor across organizational boundaries would provide a fine-grained lens through which to understand recent organizational changes in contexts where projects are central tools of innovation. Drawing on a knowledge-based perspective, some new theoretical issues are discussed, exploratory empirical evidence from the machine-tool and the global biopharmaceutical industries is provided and fertile research questions are raised for future studies.

Together with new research areas, the project-based framework proposed in the book has important managerial implications. First, as suggested in Chap. 2, sourcing decisions made across a project portfolio flexibly exploit the network of external partners (Faems et al. 2005; Vrande 2013) that can be quickly reconfigured to meet any new market changes and to handle heterogeneous technological domains. Furthermore, the advantages brought about by external sourcing can be exploited through a careful analysis of the knowledge features of the project. Second, the relationship between a multiple-project innovation strategy and multi-boundary organization strategy should be taken into consideration by firms willing to search and access external knowledge and to build new capabilities through a network of partners. The short-term plasticity of project portfolio management to quickly react to and leverage on external opportunities has to be balanced by the long-term accumulation of a firm's knowledge stock. This suggests that the opportunity to access new knowledge through the exploration activity at the project level is fully exploited when the learning-by-doing process at project level allows a firm to effectively assimilate new knowledge, overcoming cognitive barriers, and when it coherently complements the firm's technological capabilities in the long term. Third, managers should pay attention to carefully design and manage a project portfolio from the point of view of the degree of innovativeness and of the degree of openness to external partners. As discussed above and in Chap. 3 balancing different degrees of innovativeness and openness at project and multi-project level might nurture ambidexterity. Moreover, project portfolio management could be effectively implemented not only leveraging on absorptive capacity (Cohen and Levinthal 1990) in order to recognize and assimilate external new technological ideas, but also activating organizational capabilities necessary to tackle issues related to multiple partner selection, coordination, negotiation, and contractual agreement management.

Finally, the impact of project features on how firms organize their innovation processes across boundaries has been explored. However, the ways in which the external network of a firm's partners impacts on project management deserve future investigation. The exploitation of a specific group of external sources could affect a firm's decisions regarding the decomposition of innovative tasks along project stages in order to easily and quickly allocate them. For instance, the opportunity to solve a design problem by tasking a crowd of designers might inform the way managers divide innovative labor and design an incentive structure. Indeed, in the case of crowdsourcing a firm has to leverage on design expertise distributed among the crowd that can be more or less difficult to mobilize depending on the type of problem and its decomposition (Marjanovic et al. 2012).

In light of the recent research gaps discussed above, we suggest that studying the complex relationship among project features and a firm's decisions about the division of innovative labor across organizational boundaries opens some new lines for future research and contributes to the understanding of how companies implement open innovation.

References

Bahemia, H., Squire, B., (2010). A contingent perspective of open innovation in new product development projects. *International Journal of Innovation Management* 14(4): 603–627.
Besner, C., Hobbs, B., (2013). Contextualized project management practice: a cluster analysis of practices and best practices. *Project Management Journal* 44(1): 17–34.
Bonesso, S., Comacchio, A., Pizzi, C., (2011). Technology sourcing decisions in exploratory projects. *Technovation* 31(10–11): 573–585.
Brady, T., Davies, A., (2004). Building project capabilities: from exploratory to exploitative learning. *Organization Studies* 25(9): 1601–1621.
Brewer, P. A., (2007). Operationalizing psychic distance: a revised approach. *Journal of International Marketing* 15(1): 44–66.
Bruce, A.G., Martin, R.D., (1989). Leave-k-Out diagnostics for time series. *Journal of the Royal Statistical Society*. Series B (Methodological): 51(3): 363–424.
Brusoni, S., Prencipe, A., (2006). Making design rules: a multidomain perspective. *Organization Science* 17(2): 179–189.
Brusoni, S., Prencipe, A., Salter, A., (1998). Mapping and measuring innovation in project-based firms. *CoPS Working Paper* No.46, SPRU, University of Sussex.
Cassiman, B., Di Guardo, M.C., Valentini, G., (2010). Organizing links with science: cooperate or contract? A project-level analysis. *Research Policy* 39 (7): 882–892.
Cattani, G., Ferriani, S., Frederiksen, L., Taube, F., (2011). *Project-based organizing and strategic management*. Howard House, Wagon Lake, UK: Emerald Group Publishing.
Cohen, W.M., Levinthal, D.A., (1990). Absorptive capacity: a new perspective on learning and innovation. Administrative Science Quarterly 35(1): 128–152.
Christiansen, J.K., Gasparin, M., Varnes, C.J., (2013). Improving design with open innovation a flexible management technology. *Research-Technology Management* 56(2): 36–44.
Cooper, R., Edgett, S., Kleinschmidt, E., (1999). New product portfolio management: practices and performance. *Journal of Product Innovation Management* 16(4): 333–351.
Davies, A., Brady, T., (2000). Organisational capabilities and learning in complex product systems: towards repeatable solutions. *Research Policy* 29 (7–8): 931–953.
Dagenais, M.G., (1969). A threshold regression model. Econometrica 37(2): 193–203.
Dagenais, M.G., (1975). Application of a threshold regression model to household purchases of automobiles. *The Review of Economics and Statistics* 57(3): 275–285.
Dow, D., Karunaratna, A., (2006). Developing a multidimensional instrument to measure psychic distance stimuli. *Journal of International Business Studies* 37, 578–602.
Du, K.-L., (2010). Clustering: a neural network approach. *Neural Networks* 23(1): 89–107.
Evans, J., Mavondo, F.T., (2002). Psychic distance and organizational performance: an empirical examination of international ratailing operations. *Journal of International Business Studies*. 33(3): 515–532.
Faems, D., Van Looy, B., Debackere, K., (2005). Interorganizational collaboration and innovation: toward a portfolio approach. *Journal of Product Innovation Management* 22(3): 238–250.
Gardner, H.K., Gino, F., Staats, B.R., (2012). Dynamically integrating knowledge in teams: transforming resources into performance. *Academy of Management Journal* 55(4): 998–1022.
Gemünden, H.G., Killen, C., Kock, A., (2013). Implementing and informing innovation strategies through project portfolio management. *Creativity and Innovation Management* 22(1): 103–104.
Grönlund, J., Sjödin, D.R., Frishammar, J., (2010). Open innovation and the Stage-Gate process: a revised model for new product development. *California Management Review* 52(3): 106–131.

Gulati, R., Puranam, P., Thusman, M., (2012). Meta-organization design: rethinking design interorganizational and community contexts. *Strategic Management Journal* 33(6): 571–586.

Hoang, H., Rothaermel, F.T., (2010). Leveraging internal and external experience: exploration, exploitation and R&D project performance. *Strategic Management Journal* 31 (7): 734–758.

Hobday, M., (2000). The project-based organization: an ideal form for managing complex products and systems? *Research Policy* 29(7–8): 871–893.

Holzmuller, H.H: Kasper, H., (1990). The decision-maker and export activity: a cross-national comparison of the foreign orientation of Austrian managers. *Management International Review*. 30(3); 217–230.

Hsieh, K. N., Tidd, J., (2012). Open versus closed new service development: the influences of project novelty. *Technovation* 32(11): 600–608.

Huizing, E.K.R.E., (2011). Open innovation: state of the art and future perspectives. *Technovation* 31(1): 2–11.

Jerry, J., (2008). How project management office leaders facilitate cross-project learning and continuous improvement. *Project Management Journal* 39(3): 43–58.

Kimura, F., Lee, H., (2006). The gravity equation in international trade in services. *Review of World Economics* 142(1): 92–121.

Lakhani, K.R., Tushman, M.L., (2012). Open innovation and organizational boundaries: the impact of task decomposition and knowledge distribution on the locus of innovation. *HBS Working Paper* 12–057.

Laslo, Z., (2010). Project portfolio management: an integrated method for resource planning and scheduling to minimize planning/scheduling-dependent expenses. *International Journal of Project Management* 28(6): 609–618.

Lenfle, S., (2008). Exploration and project management. *International Journal of Project Management* 26(5): 469–478.

Lichtenthaler, U., (2011). Open innovation: past research, current debates, and future directions. *Academy of Management Perspectives* 25(1): 75–93.

Liu, L., Leitner D., (2012). Simultaneous pursuit of innovation and efficiency in complex engineering projects-a study of the antecedents and impacts of ambidexterity in project teams. *Project Management Journal* 43(6): 97–110.

MacCormack, A., Crandall, W., Henderson, P. and Toft, P., (2012). Do you need a new product-development strategy? *Research-Technology Management* 55(1): 34–43.

Marjanovic, S., Fry, C., Chataway, J., (2012). Crowdsourcing based business models: in search of evidence for innovation 2.0. *Science and Public Policy* 39(3): 318–332.

Mina, A., Bascavusoglu-Moreau, E., Hughes, A., (2013). Open service innovation and the firm's search for external knowledge. *Research Policy* 43(5): 853–866

Parmigiani, A., (2007). Why do firms both make and buy? An investigation of concurrent sourcing. *Strategic Management Journal* 28(3): 285–311.

Parmigiani, A., Mitchell, W., (2009). Complementarity, capabilities, and the boundaries of the firm: The impact of within-firm and inter-firm expertise on concurrent sourcing of complementary components. *Strategic Management Journal* 30(10): 1065–1091.

Perks, H., (2007). Inter-functional integration and industrial new product portfolio decision making: exploring and articulating the linkages. *Creativity and Innovation Management* 16(2): 152–164.

Picci, L., (2010). The internationalization of inventive activity: a gravity model using patent data. *Research Policy* 39(8): 1070–1081.

Powell, W.W., Koput, K.W., Smith-Doerr, L., (1996). Interorganizational collaboration and the locus of innovation: networks of learning in biotechnology. *Administrative Science Quarterly* 41(1): 116–145.

Prencipe, A., (2003). Corporate strategy and systems integration capabilities - managing networks in complex systems industries, in Prencipe, A., Davies, A., Hobday, M. (eds). *The Business of Systems Integration*. pp. 114–132, Oxford: Oxford University Press.

Reich, B.H., Gemino, A., Sauer, C., (2012). Knowledge management and project-based knowledge in it projects: a model and preliminary empirical results. *International Journal of Project Management* 30(6): 663–674.

Salge, T. O., Farchi, T., Barrett, M.I., Dopson, S., (2013). When does search openness really matter? A contingency study of health-care innovation projects. *The Journal of Product Innovation Management* 30(4): 659–676.

Söderlund, J., Vaagaasar, A.L., Andersen, E.S., (2008). Relating, reflecting and routinizing: developing project competence in cooperation with others. *International Journal of Project Management* 26(5): 517–526.

Turner, N., Maylor, H., Swart, J., 2013. Ambidexterity in managing business projects-an intellectual capital perspective. *International Journal of Managing Projects in Business* 6(2): 379–389.

Van de Vrande, V., (2013). Balancing your technology-sourcing portfolio: how sourcing mode diversity enhances innovative performance. *Strategic Management Journal* 34(5): 610–621.

Vuori, E., Mutka, S., Aaltonen, P. and Artto, K., (2013). That is not how we brought you up: how is the strategy of a project formed? *International Journal of Managing Projects in Business* 6(1): 88–105.

West, J., Salter, A., Vanhaverbeke, W., Chesbrough, H., (2014). Open innovation: The next decade. *Research Policy* 43(5): 805–811.

Wheelwright, S.C., Clark, K.B., (1992). Creating project plans to focus product development. *Harvard Business Review* March, 70–82.

Zirpoli, F., Becker, M.C., (2011). The limits of design and engineering outsourcing: performance integration and the unfullled promises of modularity. *R&D Management* 41(1): 21–43.

Editor's Biography

Sara Bonesso is Assistant Professor of Business Organization and Human Resources Management at the Department of Management at Ca' Foscari University Venezia where she received her Ph.D. in Management. She has been visiting Scholar at the Industrial Performance Center del Massachusetts Institute of Technology (Boston, USA) and at the Fraunhofer Institute for Systems and Innovation Research ISI (Karlsruhe, Germany). She is the Vice Director of the Ca' Foscari Competency Centre, a research center aimed to assess and develop the behavioral competencies in higher education contexts. She is also a member of the Teaching Committee of the Ph.D. program in Management.

She teaches in Bachelor and Master degree programs at Ca' Foscari University Venezia. Her research interests focus on antecedents of technology souring at project level analysis, absorptive capacity, technology transfer process, ambidexterity, and innovative behaviors. The research products have been presented in several international and national conferences such as EURAM (European Academy of Management), BAM (British academy of management), AOM (Academy of Management), DRUID, EGOS (European Group of Organizational Studies), IPDMC (International Product Development Management Conference), and ICEI (International Congress of Emotional Intelligence). Her recent publications include: Bonesso S., Gerli F., Scapolan A., 2014. The individual side of ambidexterity: Do individuals' perceptions match actual behaviors in reconciling the exploration and exploitation trade-off?, European Management Journal, vol. 32, pp. 392–405. Comacchio, A., Bonesso, S., 2012. Performance Evaluation for Knowledge Transfer Organizations: Best European Practices and a Conceptual Framework in Hongyi Sun, Management of Technological Innovation in Developing and Developed Countries, Reijka, Intech, pp. 127–152. Comacchio, A., Bonesso, S., Pizzi, C., 2012. Boundary spanning between industry and university: The role of technology transfer centres, The Journal of Technology Transfer, 37, 943–966. Bonesso, S., Comacchio, A., Pizzi, C., 2011. Technology sourcing decisions in exploratory projects, Technovation 31 (10–11), 573–585.

S. Bonesso et al. (eds.), *Project-Based Knowledge in Organizing Open Innovation*,
DOI: 10.1007/978-1-4471-6509-5, © Springer-Verlag London 2014

Anna Comacchio, Ph.D. in Management—Ca' Foscari University—Venezia, Italy, is Full Professor of Organization and management at the Department of Management. She has been Visiting Scholar at Centre of Corporate Strategy and Change—Warwick Business School—University of Warwick—UK. She is the Director of the PHD program in Management of Ca' Foscari University, where she teaches Organization Theory and is a member of the Board of the Graduate School of Economics and management (Ca Foscari University, Padua University and Verona University). She is Vice-Director of the Ca' Foscari Summer School. Among her institutional assignments she was the Director of the bachelor degree in Economics and Management (entirely taught in English) Ca' Foscari (2010–2013), and Director the Master in Economics and Tourism of Ca' Foscari University (2003–2008), and Rector Delegate for internationalization (2011–2013). She was the Italian delegate of the Euram board (European Academy of Management) (2009–2012). She is a member of the Scientific Committee of the Ca' Foscari Competency Centre.

Her areas of interests are organizational design and people management for innovation, innovative entrepreneurship, metaphor, and innovation. Her main focuses are absorptive capacity, knowledge transfer organizations, conceptual combination and innovation, entrepreneurship and start-up, and project-based open innovation. She participated in and coordinated several research projects on organization design and innovation. She presented several papers at international top conferences such as DRUID, AOM, BAM, (British academy of management), EURAM (European academy of management), EGOS, IPDMC (International Product Development Management Conference), she was Chair at Nobles Colloquia 2011 Venezia (Italy) and Best refereed paper presented at the eLearninternational, World conference, Edinburgh, Scotland, UK. February 9–12th, 2003. Her recent publications include: Bonesso, S., Comacchio, A., Pizzi, C., 2011. Technology sourcing decisions in exploratory projects, Technovation 31 (10–11), 573–585; Comacchio, A., Bonesso, S., 2012. Performance Evaluation for Knowledge Transfer Organizations: Best European Practices and a Conceptual Framework in Hongyi Sun, Management of Technological Innovation in Developing and Developed Countries, Reijka, Intech, pp. 127–152. Comacchio, A., Bonesso, S., Pizzi, C., 2012. Boundary spanning between industry and university: The role of technology transfer centres, The Journal of Technology Transfer, 37, 943–966.

Claudio Pizzi is Associate Professor of Economic Statistics at the Department of Economics at University Ca' Foscari Venezia since 2005. His areas of interests include linear and nonlinear time series analysis, nonlinear cointegration, evolutionary algorithms, statistical analysis of the international trade and investment, innovation, and technology transfer. He has been editor of two books published by Springer on Mathematical and Statistical Methods for Actuarial Sciences and Finance. He has participated in several research projects. He presented several papers at international conferences. His recent publications include: Bonesso, S., Comacchio, A., Pizzi, C., (2011). Technology sourcing decisions in exploratory projects, Technovation 31; Comacchio, A., Bonesso, S., Pizzi, C., (2012). Boundary spanning between industry and university: The role of technology transfer centres, The Journal of Technology

Transfer; Corò G., Giansoldati M., Pizzi C., Volpe M. (2012) The Outward Projection of Italian Firms: Facts and Figures on Foreign Affiliates, Transition Studies Review, 1/12; Gerolimetto M., Pizzi C., Procidano I. (2012) A procedure to detect hidden cointegration with the sieve bootstrap, Advances and Applications In Statistics, 30; Corazza M., Pizzi C. (eds) (2014) Mathematical and Statistical Methods for Actuarial Sciences and Finance, Springer.

Printed in the United States
By Bookmasters